Environmental Cancer—A Political Disease?

S. Robert Lichter and Stanley Rothman

YALE UNIVERSITY PRESS
NEW HAVEN
LONDON

Published with assistance from the Louis Stern Memorial Fund.

Designed by Gregg Chase.
Set in Monotype Bembo and Syntax type by Keystone Typesetting, Inc.
Printed in the United States of America by BookCrafters, Chelsea, Michigan.
Library of Congress Cataloging-in-Publication Data
Lichter, S. Robert.
Environmental cancer—a political disease? /
S. Robert Lichter and Stanley Rothman.
p. cm.
Includes bibliographical references and index.
ISBN 0-300-07306-2 (cloth : alk. paper). — ISBN 0-300-07634-7 (pbk. : alk. paper)
1. Cancer—Environmental aspects.
2. Cancer—Political aspects.
3. Environmentalism.
I. Rothman, Stanley, 1927–
II. Title. RC268.25.L53 1999 616.99'4—dc21
98-25559
A catalogue record for this book is available from the British Library.
The paper in this book meets the guidelines for permanence and durability of the Committee on Production Guidelines for Book Longevity of the Council on Library Resources.
10 9 8 7 6 5 4 3 2 1

To Sam Beer and in memory of Louis Hartz: great scholars,
great teachers,
great mentors.—Stanley Rothman

Contents

Preface

Cancer has been called the "dread disease" (Patterson, 1987). Americans fear it like no other illness. It sneaks up silently, is cruel and painful, and often results in bodily mutilation and death. Despite the huge amount of money spent on cancer since then-president Richard M. Nixon announced a war on the disease, little progress had been made toward understanding its etiology or cure until, suddenly, in the mid 1990s, a series of breakthroughs occurred in the treatment of breast cancer, and, perhaps more important in the long run, leads to possible advances in the more general treatment of the illness or illnesses. The period also witnessed, for the first time, a significant downturn in cancer rates, partially as a result of changes in American lifestyles. The scientific community has developed a new and unusual enthusiasm, though cancer specialists continue to caution against undue optimism.[1]

Even as our knowledge of the nature and causes of cancer slowly accumulates, key issues surrounding its various forms are still being debated and have become embroiled in politics. Given that most cancers are triggered by environmental factors (broadly defined), issues of public policy cannot help being involved. For example, as our ability to detect ever smaller amounts of carcinogenic materials in our environment increases, a number of questions present themselves with increasing force. What proportion of cancers have their source in our environment? Has modern industrial society produced an increase in cancer rates? What dosage of suspected carcinogens is dangerous? What methods shall we use to measure cancer potential? Are tests using mice or rats applicable to human beings? How and to what extent should we control potentially cancer-causing activities?

Views about cancer and other environmental issues influence public policy and determine how much money is spent on what. Obviously, public and private funds are not unlimited and we have to make tough choices. Monies spent on cleaning up toxic dumps are not available for AIDS research, building new schools, research on breast cancer, or Medicaid.

In the United States, the key public elite groups involved in making decisions about both funding and regulation are members of Congress, the president, judges, and bureaucrats. Other elite groups from the private sector, however, are also heavily involved in influencing what courses of action are chosen. These include industry lobbyists, environmental activists, the mass media, and the scientific community. What roles are played by these groups in influencing public opinion and hence public policy? Who determines how we spend the limited funds available to us? Who decides which scientific studies get the most public attention?

This book describes the controversy surrounding the question of environmental cancer, emphasizing the interactions among environmental activists, scientists, and the news media, as well as public officials. We want to know how the scientific community conceives of environmental cancer and its risks and to what extent scientists' findings accord with the views of mainstream environmental organizations. We also want to know how accurately the media report the views of scientific experts on environmental cancer and, in general, to what extent the consensus of the expert community is reflected in the risk assumptions underlying public policy. To answer these questions, we administered a lengthy survey to random samples of cancer researchers and top officials in leading environmental organizations. We then compared their views to those discovered in a systematic content analysis of relevant television and newspaper stories over a sufficient period (1970–1993) to trace long-term trends.

Our studies confirm what most readers probably already know—namely that different elite leadership groups in American society have radically different perceptions about the dangers of environmental cancer. In particular, leaders of environmental groups tend to think that man-made activities are a much larger threat than do businessmen and businesswomen. Other findings, however, are counterintuitive and will surprise even well-informed readers. Our findings shed further light on public policy controversies on this issue and the relation of concerns with environmental cancer to the much larger history of the growth and impact of the contemporary environmental movement in the United States and Europe.

An understanding of that history and its impact must in turn take into account the general social and political views of environmentalists and must explain why such views have so rapidly become popular among key segments of the populace. We agree with Aaron Wildavsky and Mary Douglas (1982, 1988, 1995) that the environmental movement is not merely about the environment per se. Rather, it is partly motivated by a cultural perspective on the nature of the good society. This worldview determines, to some extent at least, how environmental groups and their media allies as well as businessmen and businesswomen, political figures, and scientists view reality and how they deal with scientific evidence.

Chapters 1 and 2 trace the development of the contemporary environmental movement, including its various ideological strands and the public policies that have resulted from its initiatives. Chapter 3 summarizes the scientific literature on the environmental contribution to the etiology of "the dread disease." We are especially interested in the panics over "environmental" carcinogens in the 1970s and 1980s. The public concern, fed by the television networks, produced a "carcinogen of the month" syndrome that has yet to abate and concerning which there is still much dispute.

Chapter 4 compares the views on environmental cancer held by the environmental leadership with those of leading cancer researchers and outlines those areas in which they agree and disagree. Chapter 5 examines national media coverage of environmental cancer over the past quarter of a century. We note scientists' evaluations of the accuracy of media reporting on cancer and their views of the information dispensed by major environmental groups.

The final chapter summarizes our empirical findings and turns to broader theoretical issues. Why has the threat of man-made carcinogens been dealt with the way it has by various environmental groups and the media? Why has the scientific evidence about such matters been distorted? In light of our findings, what is the future of the environmental movement in the United States?

We conclude that environmental issues partly derive from other political concerns, and that environmental cancer is at least partly a political disease. The politics of environmental cancer is derived to some degree from traditional ideological conflicts about capitalism and the role of the state. They also, however, reflect a set of social and political orientations whose origin is more recent and whose future is unclear.

Despite the arguments of some "social constructionists," we do not believe that the views of scientific experts in this field are primarily motivated by ideological concerns, based on their social location or the source of financial support. We argue for the prudence of relying upon the consensus of the scientific community (as to the facts of the case, but not as to public values) in making policy decisions.

Finally, we examine the changes that are taking place in the environmental movement and the media. On the one hand, we find that some journalists have grown skeptical about the "green" agenda, and there are indications of comparable shifts in some mainstream environmentalist groups. On the other hand, there are still tendencies to identify new environmental dangers on the basis of less than adequate scientific evidence. (On the other side, of course, one finds a refusal to accept the adequacy of evidence about environmental dangers.) There also are signs of the increasing popularity of views that almost totally reject the modern world and its scientific underpinnings.[2]

We cover a broad range of materials and do not pretend to be equally knowledgeable in every area. We are also aware that the issues surrounding environmental cancer are changing rapidly. Indeed, change was sufficiently marked during the production of this book, thus we have been forced to modify at least some of the text to bring our materials up to date in certain key areas, though expense and time constraints have dictated limits on this.

Tapes and codebooks from our research will be deposited at the Roper Center of the University of Connecticut for use by other interested scholars one year after publication.

The study which led to this book was designed and organized by Stanley Rothman at his Center for the Study of Social and Political Change at Smith College. After many delays in completing an earlier version of the study, a revision of the survey research and the content analysis of mass-media coverage of environmental cancer was completed in collaboration with the Center for Media and Public Affairs, co-directed by S. Robert Lichter. Professor Lichter, in turn, drew upon the resources of the Statistical Assessment Service, of which he is president.

Professor William Lunch of Oregon State University is largely responsible for chapter 3 of this book but does not necessarily share the views we express elsewhere. The chapter was originally written by him some years ago but never published and has been substantially revised and brought up to date by Stanley Rothman with the assistance of Eva Isabella Celnick, a graduate student at Oregon State University, and others. Chapters 1 and 2 are based on research completed by Professor Amy Black of Franklin and Marshall College, as revised by Rothman, with the same caveat.

We thank Michael Gough, Harvey Brooks, Paul Gross, Martin Lewis, and Donald Baumer for reading an earlier version of the manuscript and making valuable suggestions. We also thank Elizabeth Whelan for bibliographic assistance. John Tolman helped with updating the material before and during the copyediting of the manuscript. He also caught some important errors and even rewrote the section on tobacco in chapter 3. Naturally, the errors that remain are those of the authors.

A final thank-you goes to Jinny Mason, the indispensable Jill-of-all-trades assistant for the Center for the Study of Social and Political Change.

The study was supported by grants from the Sarah Scaife Foundation, the Olin Foundation, and the Earhart Foundation.

I
Historical Lessons of the Environmental Movement

THREE DECADES AGO, FEW AMERICANS could have imagined the extent to which environmental issues now pervade their society. Everything appears to be turning "green." Manufacturers proudly advertise products packaged with recycled materials; recycling bins are appearing in airports and schools, on neighborhood curbs, and in grocery stores. Entire lines of environmentally friendly and recycled products, from paper goods to cleaning supplies to greeting cards, line store shelves. Schools are incorporating ecological issues into the curriculum; even children's cartoons and toys reflect environmental themes.

Many politicians actively compete for the green vote, wooing voters with promises to support policies of environmental protection. State referenda address such issues as recycling, offshore drilling, and endangered species. The apparent explosion of environmentally friendly politics, education, and consumerism is just one of the many recent manifestations of the rapid growth of environmentalism in the United States. The Republican congressional victory of 1994 seemed to mark a backlash, but most of the initiatives pushed by the Republican leadership failed. Indeed, President Bill Clinton has used the Republican stance on the environment to bolster his own strength, and the Republicans have curbed their attack on environmental programs since the 1996 election. To understand the surge of interest in environmental issues since the mid-1960s, and to understand the state of environmentalism in the United States today, we must first examine the origins and evolution of American ecological concerns.

In the first centuries of European settlement of North America, environmental degradation was not a matter of public attention. Notwithstanding a few laws and practices, such as William Penn's 1681 ordinance requiring that the people of Pennsylvania "leave an acre of trees for every five acres cleared" (Nicholson 1987, p. 26), the predominant concern of early settlers was to secure the basic necessities of life, to build homes, to farm the land. The historian Hans Huth, noting the massive clearing of forests to create farmland, describes the early American experience as one devoted to progress and expansion, often at the expense of the environment. Travelers' accounts, songs, and poetry from the era reflect the ethos of expansion: the axe came to symbolize the relationship between colonists and the natural environment (Huth, 1957, p. 2).

To most early Americans, the frontier was endless, and natural resources were so bountiful as to seem inexhaustible. Unlike densely populated Europe, North America offered vast expanses of land to those willing to venture westward. But a number of dramatic changes in the eighteenth and nineteenth centuries transformed American society and facilitated the development of an increasingly widespread concern for nature. The formation of an independent United States and the Revolutionary War changed the political order. Although the new national government under the Articles of Confederation was weak in comparison with that of most states, the adoption of the Constitution in 1789 was a significant nationalizing force that created the opportunity for the growth of a powerful federal government. Over the course of the next two centuries, the scope of national laws expanded to encompass almost every aspect of American life. Issues formerly relegated to state governments or left for private citizens to settle became national matters, and the federal government became a powerful forum for addressing environmental problems.

Just as political change fostered the development of nationwide political movements, so did the Industrial Revolution forever change American life. Although the advances of industrialization resulted in many positive societal changes, the negative consequences of industrialization posed serious environmental threats—depleting natural resources, polluting the air and water, and overpopulating the land.

Historical Development of Ecological Concerns

The ideological roots of the environmental movement in the United States date back at least to 1864, when George Perkins Marsh published *Man and Nature,* the first English-language publication to detail the link between human action and environmental degradation.[1] Marsh presented a compelling argument that destructive practices, often intended to further civilization, were actually destroying the earth and threatening the extinction of human beings (Marsh 1965 [1864], 39–40). Noting the detrimental ecological effects throughout history of practices such as irrigation and deforestation, Marsh argued that men and women were ethically bound to care for the earth.

Marsh's work provided intellectual underpinnings for the modern ecological movement. Although scientists today reject Marsh's view of nature as a steady-state system that changes only when disturbed by human intervention, that concept has been a fundamental premise of modern environmental laws and treaties, ecological theory, and ecological writings for more than a century (Botkin 1990, pp. 9–10). The flaw of the steady-state hypothesis does not, however, invalidate Marsh's principal thesis that human actions can degrade the environment. His writings provided a sober warning for proponents of technological change: humans often destroy nature in the name of progress.

Although Marsh's work focused on ecological issues, the term *ecology* was first used by Ernst Haeckel in 1866 to describe "the science of relations between organisms and their environment" (Bramwell 1989, p. 40). Haeckel saw ecology as a tool for understanding the interrelationships between biology and geography. His works influenced other scientists, who developed and expanded his conception of ecology to incorporate all the life sciences.

In Europe and the United States in the latter half of the nineteenth century, there developed two distinctive perspectives on nature—preservation and conservation. Although proponents of these outlooks differed in their views of the purpose and value of nature, they were united in their concern for the environment.

Preservationists, influenced by the Romantic conception of nature as a source of inspiration, opposed any use of wilderness lands except for recreation. Emerson and Thoreau, among the transcendentalists, embraced a naturalist philosophy that emphasized the connection between spirituality and nature. Central tenets of transcendentalism included the beliefs that the soul can transcend normal understanding to grasp spiritual truths and that nature reflects the divine (Nash 1982, p. 85). According to this perspective, returning to the wilderness would reduce the destructive forces of civilization; by seeking the "Romantic value of nature as the ultimate restorer and purifier of humanity corrupted by civilization" (Petulla 1977, p. 228), men could renew their faith and find truth.

Perhaps the person who best typifies the early preservation movement in the United States is the philosopher and nature writer Henry David Thoreau. Centering on the interconnection of man and nature, Thoreau's works chronicle the evolution of his appreciation of the wilderness.[2] Reflecting his belief in the innate goodness of mankind and his distrust of civilization, Thoreau saw wilderness as a source of inspiration and renewal. In his later writings, he advocated seeking a balance between the wild and the civilized (Nash 1982, pp. 90–92). Thoreau's widely read works continue to influence environmental thought to the present day. As one historian concludes, "It is no exaggeration to say that today all thought of the wilderness flows in *Walden*'s wake" (Oelschlaeger 1991, p. 171).

Whereas Thoreau preferred to experience nature in areas not far from civilization, one of the earliest American environmental activists, John Muir, explored and understood the remotest depths of wilderness. Muir devoted his life to the fight for preservation, describing his outdoor adventures and promoting his wilderness philosophy in almost 150 articles in *Harper's, Century,* and other popular magazines (Whitaker 1976, p. 20). An avid believer in the protection of wilderness and the promotion of natural recreation, Muir was a vocal advocate for the establishment of forest reserves. In the late 1890s, he founded the Sierra Club to promote the recreational use and enjoyment of the mountain regions of the Pacific coast (Petulla 1977, pp. 231–233). Through the course of his experiences in the wilderness, Muir developed a biocentric, pantheistic worldview, finding the divine in all nature and rejecting the anthropocentric order (Oelschlaeger 1991, pp. 180–181).

For all his idealism and good works, Muir's preservationism was based on a series of fallacies. There was, by his time, no pure nature left to preserve, though that was not as clear in the United States as it was in Europe. The migration of the human species to North America appears to have resulted in the extinction of numerous species of animals. Sections of the grassland Great Plains had been formed by Amerindians clearing forests, and the Amerindians on horseback who met European settlers of the American West in the mid-1800s had developed their way of life only about a hundred years earlier, when the Spanish conquistadors introduced horses (Lewis 1992; Easterbrook 1995; Budiansky 1995). In any event, the so-called natural environment that the Europeans encountered dated only from the end of the last Ice Age.

While Muir and other preservationists campaigned to protect the earth from human domination, proponents of an alternative environmental perspective gained influence. Resource conservationism, the scientific movement of early environmentalism, emphasized efficient use of the land and its natural resources. Inspired by the Enlightenment's veneration of progress, conservationists valued advancement and signs of "civility." Although conservationists shared preservationists' concerns about human destruction of nature, they believed scientific progress would lead to solutions to ecological problems.

The official closing of the American frontier, announced in the census of 1890 and popularized in the writings of Frederick Jackson Turner, provided an impetus for the conservation movement. For the first time Americans had to confront the reality of limited natural resources, and conservation offered a means for developing solutions (Whitaker 1976, p. 17). Interested in introducing farming techniques such as crop rotation, pest control, and fertilization, which sought to conserve and enrich land resources, adherents of this scientific school established academic journals, agricultural societies, and other scientific organizations to promote innovation and learning (Petulla 1977, pp. 218–219).

Even though modern discussions of preservation and conservation distinguish between these two streams of early ecological thought, their contemporaries would not have made such a distinction.[3] For most of the nineteenth century, proponents of preservation and conservation shared many common goals and avoided direct confrontations.

The national park movement is an example of an early environmental effort that received support from advocates of both viewpoints. First proposed in 1832 by the artist George Catlin as a way of preserving wilderness (Nash 1976, p. 5), the idea of national parks was furthered by Frederick Law Olmsted, the landscape architect who designed New York's Central Park. Olmsted played a pivotal role in convincing the government to preserve Yosemite Valley (Petulla 1977, p. 230). In 1864, Congress designated Yosemite Valley protected for "public use, resort, and recreation," naming the state of California trustee (Nash 1976, p. 18). In 1872, Congress set aside two million acres to create Yellowstone National Park, the first national park in the world. Carl Shurz, Secretary of the Interior in the 1870s, furthered the cause of national parks, beginning an assault on deforestation that, among other things, called for the creation of forest reserves (Udall 1988, pp. 86–87). Over the next four decades, the United States government added millions of acres to the national park system.[4]

In the late 1800s, a number of significant social and political shifts provided the impetus for a national conservation movement. Industrialization brought poverty, crime, pollution, and disease as urban areas grew more crowded. In newspapers muckrakers exposed hazards in the workplace and corruption in government. Concerns about resource depletion escalated with the closing of the frontier.

Fueled by a perception of the worsening environmental conditions and government apathy, protest groups formed to demand political change. Local citizen organizations devoted to environmental issues raised public health concerns, calling for action against pollution and demanding improvement of municipal services (Melosi 1992, pp. 100–101). The growing dissatisfaction with the quality of life in America created trends that significantly furthered the environmental cause. The rise of the "wilderness cult," whose adherents upheld the value of preserving wilderness as an alternative to civilized society, coincided with the trend toward greater public participation in government (Nash 1982, pp. 143–145; Douglas and Wildavsky 1982, p. 155). The Progressive movement, in large part a reaction to the growing discontent, advocated increased citizen participation and promised greater governmental responsiveness to citizens' concerns.

Increased environmental concern and a desire for political change culminated in the Progressive Conservation Movement of 1890–1920, an effort primarily focused on ensuring the sustainable use of land, forests, and water (Petulla 1977, p. 267). Championed by Theodore Roosevelt, the Conservation Movement politicized environmental concerns and marked the beginning of federal environmental policy (McConnell 1954, p. 464). Progressive conservation claimed to speak for the people, promoting the preservation of public lands and implementation of environmental policies to secure a sustainable level of natural resources.

Because the utilitarian aims of Progressive conservationists fundamentally differed from the goals of preservation-minded activists, tensions between the two camps escalated, culminating in debate over the proper use of public lands. Gifford Pinchot, the head of the Department of Forestry within Roosevelt's Department of Agriculture, oversaw the rapid enlargement of the national park system. Historians consider Pinchot to be the father of the Conservation Movement. During his tenure in office, the United States added 134 million acres of national forests and began to regulate the use and conservation of national resources (Whitaker 1976, pp. 18–19). Pinchot claimed that conservation rested upon three principles: development of natural resources, prevention of wasteful destruction of the earth, and protecton of resources for the common good (McConnell 1954, p. 466). In marked contrast to Muir and the preservationists, who advocated preserving public lands for wildlife and recreation, Pinchot's philosophy centered on wise commercial use of resources on public lands. Thus, though conservationists and preservationists might agree on some practical issues—the preservation of wilderness, for example—their ultimate goals differed and could lead to violent confrontation. The conservationists wished to render the most effective use of the land for human purposes, which could, in their eyes, lead to improving nature. The preservationists, on the other hand, did not believe in such improvement. Their goal was to protect as much of the pristine wilderness as was humanly possible.

The land-use controversy culminated in the debate over the use of Hetch Hetchy reservoir, an issue that resulted in a sharp break between conservationists and preservationists (Tucker 1982, p. 55). Pinchot and resource conservationists wanted to construct a dam in the Hetch Hetchy Valley of Yosemite National Park to provide hydroelectric power and a reservoir for San Francisco. Muir and other preservationists opposed the project, arguing that national parks should be protected wilderness areas (Nash 1982, pp. 161–163). In what became a battle of economics versus aesthetics, supporters of the cost-conserving Hetch Hetchy project triumphed, and construction of the dam began. The Hetch Hetchy project created a permanent rift between conservationists and preservationists and marked the beginning of the intense battle between resource utilization and preservation (Udall 1988, pp. 120–121).[5]

In Europe, developments took a rather different turn. It was difficult to preserve a nature which had clearly been worked over by human beings since time immemorial (Dubos 1980). To European romantics, preserving nature meant preserving (or going back to) an idealized vision of the peasant fields and villages of the Middle Ages. Like contemporary environmentalists, the romantics favored organic farming, opposed experimentation with animals, and believed in vegetarianism. Ironically, some features of the European environmental movement were adopted by the Nazi party; Hitler and Himmler exemplified many of these trends, including a concern with animal rights (Bramwell 1989; Ferry 1995).[6] Some on the left were attracted to environmental ideas, but they were a relatively minor theme in Marxist and neo-Marxist political ideas and movements of the period. Most Marxists supported industrialization and looked forward to the domestication of nature by industry.

The Emergence of the Contemporary Environmental Movement

Although the Conservation Movement represented a dramatic shift of popular interest toward issues of environmental concern, the modern environmental movement, actually a fusion of public health and preservation concerns, did not emerge until the 1960s. Competing scholarly explanations of its growth stress the impact of socioeconomic, political, demographic, and technological changes, ideological shifts, and ground-breaking works and events. The one point on which various authors agree is that the rapid growth and intensity of focus on ecological issues were nothing short of revolutionary.

Survey data best demonstrate this transformation of public opinion in the late 1960s. The first opinion polls to ask questions concerning the environment appeared in 1965, gauging public concern about pollution. That year only 28 percent and 35 percent of those polled considered air and water pollution, respectively, to be serious or very serious problems. By 1970, 69 percent and 74 percent, respectively, listed air and water pollution as serious concerns. Similarly, the proportion of Americans listing ecology as the most important problem facing the country rose from only 1 percent in 1969 to 25 percent in 1971 (Whitaker 1976, p. 8). In 1970 *Time* magazine named the environment the "issue of the year," and *Life* magazine declared the 1970s the "environmental decade."

Existing accounts of the development of the contemporary environmental movement characterize it in diverse and often conflicting ways. Merril Eisenbud (1978) traced the convergence of public health and conservation concerns, arguing that the environmental movement has distracted policymakers and the public from focusing attention on improving public health. The anthropologist Mary Douglas and the political scientist Aaron Wildavsky (1982) analyzed social and political constructs that influence how Americans perceive the environment. In their view, what is perceived as pollution and increasing risk from environmental degradation is determined by cultural constructs created by cultural leaders and is not simply a function of reality. William Tucker (1982) characterized the leaders of the environmental movement as privileged members of society who seek to protect an "environment" only they can afford to enjoy. Historian Samuel P. Hays (1987) placed the movement in historical context, arguing that social and political forces of the post–World War II era shaped the development of environmental policies and issues. Henry Caulfield (1989) highlighted the significance of elite behavior in shaping and advancing the conservation and environmental movements.

Although all these authors are influenced by their disciplinary training, life experiences, and ideology, each account captures an aspect of the truth. Modern environmentalism includes numerous factions representing different ideological approaches to key issues. It follows that the development of the movement has been complex and multifaceted. Because of this complexity, no single issue or historical development alone can account for the emergence and rapid growth of the movement.

Despite the large body of literature that chronicles the development of the movement, no author has yet to articulate a complete causal explanation for the rise of contemporary environmentalism. The complexity of issues, the overlap of contributing factors, and the general difficulty of determining directions of causality have made such modeling difficult, if not impossible. We can, however, identify the significant contributing factors that provided the impetus for environmental activism.

One factor commonly cited by historians of the environmental movement is rapid socio-economic change. The increase in the national standard of living accelerated the development of modern environmentalism. The postwar economic boom created an era of growing affluence, as the middle class grew and a new mass elite developed. Improved working conditions, higher wages, and shorter work weeks gave workers more free time and extra money to enjoy recreation (Whitaker 1976, p. 24).

Concurrent with the rising standard of living was an increase in consumerism. The American conception of the necessities of life changed. No longer content with just shelter, clothing, and food, the minimum standard of living was redefined to include durable and material goods (Hays 1987, p. 35). Consumerism became the hallmark of American life: only 6 percent of the world's population lived in the United States in the 1950s, yet Americans consumed more than a third of the world's goods and services (McCormick 1989, p. 50).

American consumerism has a cultural context. In *People of Plenty,* David Potter traces the historical roots of American materialism. Drawing heavily on the behavioral sciences, Potter describes Americans as socialized, even from infancy, into a culture of abundance (Potter 1954, pp. 190–208). A seemingly inexhaustible supply of resources promised by an open frontier was quickly replaced with the promise of industrialization and new technologies, renewing the sense of economic abundance (Potter 1954, pp. 164–165). According to Potter's thesis, the rapid consumerism that developed in the era of affluence following World War II is a natural development, given preexisting social and cultural trends.

Many commentators have noted the connection between concern for the environment and socio-economic status, arguing convincingly that as the standard of living improves, concern for the environment grows. When people are delivered from worrying about economic issues, they can turn their attention to improving the overall quality of life they enjoy. In the postwar era, rising levels of economic prosperity freed many Americans to concern themselves with quality-of-life issues and created an ideal climate for challenging the status quo (Hays 1987; Inglehart 1977, 1990, 1995).

Almost all accounts ascribe importance to the unique social conditions and political climate of the 1960s. The nuclear arms race, rising racial tensions, and military action in Southeast Asia sparked political upheaval. Citizens concerned about both foreign and domestic issues organized to fight against policies they opposed.

The Civil Rights movement was the first postwar movement of significant scope that challenged the established social order. The antiwar movement echoed many of the same antiestablishment themes among upper-middle-class whites. Advocates of "counterculturalism" rebelled against the fundamental tenets of the dominant social order, calling for peace, freedom, and love uninhibited by cultural, social, or political structures (McCormick 1989, pp. 61–64).

A bond of antimaterialism united most of the upper-middle-class whites participating in various movements (Hays 1959, p. 145). Another common thread was the increased involvement of students in protest movements. Beginning with civil rights demonstrations and escalating with the predominantly campus-centered antiwar protests, students' political energies eventually were channeled into the ecological movement (Nash 1982, pp. 252–253).

The political changes of the 1960s opened the door for nationwide grassroots environmental activism. As the momentum of earlier social movements waned, activists found new causes to champion, including environmentalism. In a natural progression, many protesters and organizers, trained in the antiwar, antinuclear, civil-rights, and counter-cultural movements, applied the techniques and strategies learned in their earlier crusades to the rapidly expanding movement to "save the earth." This tendency is apparent in the transformation of older environmental groups, such as the Sierra Club, by urban middle-class intellectuals. Previously, the members had been conservative and upper class and interested primarily in recreational options (Faich and Gale 1971). It is also clear that the newer environmental groups were originally launched by individuals with generally radical perspectives, for whom environmentalism served as a way of attacking a society from which they were alienated. Indeed, many champions of environmentalism (such as Barry Commoner, Jeremy Rifkin, and David Dillinger) had supported a variety of radical causes in the past (Dunlap and Olsen 1984, Milbrath 1984, Gale 1983; Rubin 1994).

Rapid technological developments also facilitated the growth of the environmental movement. These developments have transformed American society to such an extent that technological advancements have become "as much a human right as life and liberty" (Bronowski 1974, p. 52).

Americans have developed a great taste for modern luxuries, but the technological developments that make their consumption possible create significant environmental problems. For example, it is difficult to overestimate the dramatic impact of the automobile on American life and environmental quality. Lewis Mumford refers to Americans as adherents to the "religion of the car" (1953), describing how they are willing to sacrifice efficiency, air quality, and aesthetic considerations for the conveniences cars provide. As the intensity of dependence on modern conveniences increases, the environmental costs generated by the use of advanced technologies are becoming more visible.[7] Scientific advances increased our knowledge about the environment and provided volumes of new data for analysis. Improved instrumentation allowed greater scientific certainty. With the help of advanced new technologies, scientists could more readily detect microscopic trace elements of chemicals. Additionally, the development of international organizations promoted cooperation among scientists worldwide and facilitated the sharing of important new discoveries.

New and improved forms of mass communication dramatically affected the political, social, and cultural climate. The development of television brought news reports and corresponding visual images into the homes of millions of Americans. News of social unrest invigorated social movements and nationalized debates over social and political issues. Similarly, advanced techniques for print media increased public awareness of ideas, places, and issues (Rothman 1992, pp. 37–74). The development and expansion of color photography provided vivid visual images useful in promoting an appreciation of nature. Environmental groups, particularly the Sierra Club and Friends of the Earth, published color photography books that introduced readers to the natural beauty of the wilderness (Hays 1987, p. 37).

The combination of increased scientific knowledge and improved forms of mass communication gave environmental groups momentum: Whitaker (1976, p. 25) observes that "vastly more was known and noted about pollution in the United States than in any other country, and this in itself provided a climate in which support for environmentally related causes could be elicited." Activists learned how to promote their views through different forms of media, formulating strategies to use televised and print images to further their goals.

Technological advances and corresponding increases in living standards resulted in demographic changes as well. With advances in medicine and rising affluence, average life expectancy increased dramatically, from an average of forty-seven years in 1900 to seventy-one years in 1970. Fewer people died in infancy or childhood, and adults lived longer as new medicines provided cures for previously life-threatening illnesses such as pneumonia, tuberculosis, and influenza. As greater numbers of people lived longer, the incidence of cancer, heart disease, and stroke—diseases often linked to or exacerbated by behavior patterns—increased (Eisenbud 1978, pp. 362–363). As death rates and infant mortality rates fell, birth rates initially remained steady, resulting in worldwide population increases (Gomer 1974, p. 64).

Aided by the mobility provided by automobiles, urban sprawl expanded as city dwellers moved out to suburbs. Between 1950 and 1959, for example, American cities grew by only 1 percent, while the population in suburbs swelled by 44 percent (Udall 1988, p. 161). Americans were growing more concerned about the quality of the environment in which they lived; the suburbs promised clean air, open spaces, and closer contact with nature. Wilderness became symbolic of true beauty and worthy of esteem (Hays 1987, pp. 23–24). Heightened concern with nature manifested itself in the movement out of cities and in the increased emphasis on the protection of wilderness and species.

All accounts of the American environmental movement agree that the works of such key writers as Rachel Carson, Barry Commoner, and Paul Ehrlich were instrumental in bringing environmental issues and themes to public awareness. Charles Rubin calls these authors "popularizers," Edith Efron uses the term "apocalyptics"; some laud them as prophets and others take them to task for playing loose with the facts. However one chooses to describe them, there is no question that they provided a "much needed environmental awakening" (Eisenbud 1978, p. 55).

Rachel Carson's 1962 book *Silent Spring* played a pivotal role in transforming public awareness of environmental issues.[8] Roderick Nash compares the impact of Carson's work with that of *Uncle Tom's Cabin,* which energized the abolition movement more than a century before (Nash 1989, p. 78). The book's title alludes to the opening chapter, "A Fable for Tomorrow," in which Carson describes a world so devastated by the secondary effects of pesticides that silence replaces the songs of birds in spring. The remainder of the book is a synthesis of previous research about the deleterious effects of chemical pesticides.[9] Carson describes the extent of risks associated with synthetic pesticides, discussing findings that dangerous chemicals such as DDT enter the food chain and ultimately threaten human health. Accentuating the interconnection of all life on earth, Carson concludes by calling for the use of "natural" remedies to control what we call "pests" and by recommending government regulation of the use of harmful chemicals.

Silent Spring was not the first work to discuss the hazards of pesticide use, but it was the first to have a significant public impact. Originally published in serial form in the *New Yorker,* Carson's work appealed to a wide audience. A talented writer, Carson explained arcane scientific studies in lively, readable prose. Although a best-seller, the influence of *Silent Spring* extended far beyond those who read the book; Carson's thesis sparked controversy and became a common topic of conversation and public concern (Nash 1989, p. 81). In the years following its publication, Congress passed legislation regulating or banning the substances Carson warned were most hazardous (Fox 1981, p. 298).

Critics contend that Carson exaggerated the harmful effects of pesticides. Although many reviews of *Silent Spring* were unequivocally favorable, some reviewers, including those for *Science,* and *Scientific American,* criticized Carson for presenting a one-sided view of pesticide use (Whelan 1992, pp. 98–99). Carson was correct, the critics argued, to alert the public to the possible hazards of the overuse and misuse of DDT. Yet she failed to discuss the incredible health benefits of pesticides in general, and DDT specifically. By controlling insect-borne diseases, DDT unquestionably saved millions of lives. For example, before DDT spraying began, Sri Lanka (then Ceylon) reported 2.8 million cases of malaria. The disease all but disappeared as a result of the use of DDT, reaching a low of seventeen cases in 1962, but its incidence escalated again when the country banned the use of the pesticide eventually exceeding preban incidence levels. By the mid-1970s, the international community had followed the United States in banning DDT, and worldwide health statistics demonstrated that malaria was making a strong comeback in a number of other countries (Ray and Guzzo 1990, p. 69; Whelan 1992, p. 101).

Boosted by what Whelan calls the "silent springboard" of Carson's work, the environmental movement organized around the issue of pesticides. Many environmental hazards that opponents of

DDT cited to justify a DDT ban have never been scientifically proven. Indeed, the National Cancer Institute concluded, after extensive testing, that DDT is not a carcinogen (Ray 1990, p. 73). Summarizing the scientific findings of numerous studies of DDT, Whelan concluded that "DDT, when used as intended, did not threaten wildlife or the environment and did not cause human disease. DDT prevented more human death and disease than any other manmade chemical in all of recorded history" (Whelan 1992, pp. 99–100).[10]

Overpopulation was the subject of many books during the 1960s, but none had the impact or popular appeal of Paul Ehrlich's *The Population Bomb*. Published in 1968, Ehrlich's book warned of an impending population crisis that threatened to create worldwide famine and destruction. Advocating zero population growth in the short run, with a long-term vision of negative population growth, Ehrlich believed that a dramatic shift in attitudes was the only hope for preventing mass starvation.

A central theme of his work was controlling reproduction. Ehrlich is a strong advocate of birth control, admitting that governments might need to resort to coercive measures to inhibit reproduction. His proposed solutions reflect his deep pessimism: they range from imposing a luxury tax on items related to childbearing to limiting agricultural aid for nations with the potential to become self-sufficient food producers.

A showman as well as a scientist, Ehrlich appeared on television promoting his ideas and his book. After his well-received appearance on *The Tonight Show,* sales of *The Population Bomb* soared (Rubin 1994, p. 77). The best-selling ecology book of the 1960s, it eventually sold more than three million copies (Udall 1988, p. 239).

Trained as an entomologist, Ehrlich's only previous book was *How to Know the Butterflies* (Udall 1988, pp. 239–240). When he became interested in the ecological effects of overpopulation, he applied his understanding of insect population growth to human population growth. His scientific background proved useful when writing about environmental issues. Yet flawed inferences that failed to account for the differences in population patterns of humans and insects led to errors in his arguments.

In the first edition of *The Population Bomb,* Ehrlich made drastic predictions that he had to revise in later editions.[11] Forecasting monumental famines by the close of the following decade, he argued: "At this late date nothing can prevent a substantial increase in the world death rate" (Ehrlich 1968, p. 11). This prediction proved false: the world did not succumb to starvation along the scale of Ehrlich's warnings, and death rates continued to decrease worldwide (Rubin 1994, p. 91). Although Ehrlich's alarming prophesies were highly exaggerated, the book incited widespread public concern about overpopulation and enlivened public debate.

The biologist Barry Commoner provided a competing explanation of the roots of an impending ecological crisis in *The Closing Circle* (1971). Labeled the "Paul Revere of ecology" by *Time* magazine, Commoner was already a vocal member of the early environmental movement when he published his ground-breaking work (McCormick 1989, p. 70). Educated in zoology and biology, he began to identify himself as an ecologist through his work protesting atomic testing in the 1950s. Commoner and his colleagues researched the environmental effects of radiation and began a public education campaign focusing on the potential hazardous effects of nuclear testing (Udall 1988, pp. 236–237; Commoner 1971, pp. 49–65).

The Closing Circle introduces Commoner's four informal "laws" of ecology which, if violated, threaten the balance of the ecosystem: (1) everything is connected to everything else, (2) everything must go somewhere, (3) nature knows best, and (4) there is no such thing as a free lunch (Commoner 1971, pp. 32–46). He demonstrates the results of violating these laws through three case studies of technological destruction of the natural order: smog in Los Angeles, nitrate pollution in the agricultural area of Decatur, Illinois, and industrial pollution "killing" Lake Erie.

The Closing Circle was partially a rebuttal to Ehrlich's *Population Bomb* (Paehlke 1989, p. 58).[12] Less concerned about the effects of overpopulation, Commoner blames technology for the tremendous increase in pollution (and subsequent ecological destruction) in the postwar era. He focuses his critique on economic causes, tracing the root of ecological problems to the capitalist goal of profit maximization. In pursuit of high profits, Commoner explains, industries attempt to increase productivity by exploiting new technologies without weighing the environmental costs. Commoner criticizes the capitalist system for not taking into account the depletion of "biological capital," the environmental cost of producing goods and services.

Although Commoner articulated his concerns about the dangers of capitalism in *The Closing Circle,* it was not until the publication of *The Poverty of Power* in 1976 that he offered explicit political solutions to the ecological crisis. Commoner argued that only government action could control the pollution created by the avarice that capitalism encouraged. Thus he maintained that socialism offers the best solution to environmental problems. In 1980, he ran as the presidential candidate for the Citizens party, a leftist organization that promoted his political and environmental agenda (Efron 1984, p. 37).

In a fourth ground-breaking study, *The Limits to Growth* (1972),[13] a group of researchers at MIT used computer modeling to predict when the present supply of natural resources would be exhausted. Their model analyzed five factors that contribute to global problems—industrialization, population growth, agricultural production, resource consumption, and pollution—and predicted that, at current rates, "the limits to growth will be reached sometime within the next one hundred years" (Meadows et al., 1972, p. 23). The study reiterated neo-Malthusian themes, warning that current rates of consumption and growth were leading to an impending crisis of overpopulation, malnutrition and starvation, resource depletion, and environmental destruction. The authors argued that only dramatic political and social changes could save the world from destruction. Calling for a "global equilibrium," Meadows and his colleagues demanded an end to economic and population growth.

The Club of Rome, an organization created and funded by the Italian businessman Aurelio Peccei to promote the search for solutions to global problems, sponsored the research reported in *Limits* (Coffman 1994, p. 103). In what Charles Rubin describes as "a textbook example of doing science by press release" (1994, p. 132), the Club of Rome orchestrated a series of events to publicize the study even before its official publication. A public relations firm distributed press releases, the Xerox Corporation funded a symposium at the Smithsonian Institution to discuss issues raised in the study, and mass media announced the shocking findings.

Limits to Growth was subject to severe methodological critique, as commentators condemned the methods of data collection, the technological soundness of the study, and the rigor of the simulation model.[14] Despite these and other criticisms, and perhaps stirred in part by the public controversy, *Limits* was a best-seller that raised public consciousness of some of the detrimental effects of economic and population growth.

Influential Events and Developments

A series of scientific advancements and developments within the environmental movement also helped generate increased public awareness and political action. Aided by improved technologies that helped scientists identify smaller and smaller amounts of potential carcinogens or other potential health hazards, many environmental groups placed increasing emphasis on the connection between public health and ecology. Media reports about new scientific findings on the environmental causes of illnesses raised public concerns about their health and safety. Fear of the harmful effects of pollution escalated, and growing numbers of citizens began demanding that the government take steps to ensure better air quality, safe water supplies, and clean waterways (Hays 1987, pp. 24–25).

Due in part to rising standards of living and increased life span, public expectations about the quality of life climbed. People expected to live long, productive lives, and they wanted to ensure their well-being as they aged (Hays 1987, p. 26). The popular media provided information about extending the quality of life, focusing public attention on the environmental causes of poor health.

On a number of occasions media attention to the link between environment and disease produced what can only be called a public frenzy. The greatest public scare was over carcinogens in the environment. As we shall argue, public information regarding the relation between health and the environment has been vastly distorted. Although in many cases pollution, pesticides, and other toxic chemicals do present health hazards, the question is the extent and severity of the risks.

No single event had a greater impact on the development of modern environmentalism than Earth Day, April 22, 1970. Heralded as "a red letter day, a coming of age for the American environmental movement" (Udall 1988, p. 244) that succeeded in "finally waking up America to this impending crisis" (Coffman 1994, p. 275), the event was a triumph of nationwide grassroots politics. Conceived by Senator Gaylord Nelson, a Democrat from Wisconsin, Earth Day was a nationwide celebration of environmentalism. With the aid of federal funding and a small team of student organizers, millions of Americans participated in rallies and teach-ins nationwide. More than fifteen hundred colleges and ten thousand schools celebrated Earth Day; a two-hour parade down New York's Fifth Avenue drew over one hundred thousand participants. Both houses of Congress recessed for the day, as Senator Nelson and others traveled across the country giving speeches. For those who did not directly participate, all major media outlets covered the day's activities, bringing news of environmental issues directly into American homes (Fox 1981, pp. 325–326; McCormick 1989, p. 67).

Earth Day was more than an occasion of rallies and media attention; it provided the impetus for a nationwide explosion of environmental concerns that had been growing for the previous decade.[15] The massive scale of citizen participation in Earth Day activities demonstrated the breadth of environmental concerns, mobilizing a mass public and convincing many legislators that they could no longer ignore environmental issues.

Samuel Hays is certainly correct in calling the 1960s and 1970s "strong and expansive years" for environmental organizations (Hays 1987, p. 53). In the first half of the twentieth century, an average of three new environmental groups formed each year. The growth surged dramatically between 1961 and 1980 to an average of eighteen new environmental public-interest organizations a year (Scheffer 1991, p. 113).

Most of these organizations were formed to represent specialized interests. While many new groups rallied support for particular areas of ecological concern, the character of existing conservation groups also underwent dramatic changes, expanding the scope and intensity of their work. The Sierra Club, for example, transformed itself from a small group of nature lovers concerned with outdoor activities in Western mountain ranges into a large-scale, powerful professional lobbying organization concerned with a wide range of environmental issues. By 1994 the Sierra Club boasted 650,000 members, a $35 million annual budget, and twenty-seven policy committees representing a broad spectrum of issues.[16]

Until the early 1960s, most environmental groups addressed very specific issues with correspondingly specific, and usually narrow, consequences. The causes of these problems were unambiguous, and the costs to solve them were often minimal. During the 1960s, however, environmental groups began addressing the more complex "second-generation" issues. Less specific, with greater scope and more significant consequences, the new set of issues usually related to public health concerns. The causes of these environmental hazards were ambiguous and difficult to prove, and solutions could be expensive. With the massive increases in scope and scale of the second-generation issues, opposition to environmental proposals increased commensurately. No longer localized concerns, the new larger-scale issues threatened entire industries and behavioral patterns (Mitchell 1991, pp. 84–88).

As environmental groups expanded, most organizations also added professional staffs to coordinate the needs of the growing membership and the widening scope of activities. In a significant departure from the more conservation-oriented groups of earlier decades, many new and renewed organizations began to employ professional lobbyists and based their operations in or near Washington, D.C. (Gottlieb 1991, pp. 45–46; Douglas and Wildavsky 1982, p. 129). As of 1994, all but three of the ten most influential environmental organizations had headquarters in Washington.

As new groups formed, each sought to set itself apart by appealing to a particular policy or ideological concern. Capitalizing on the spillover of political activism from the 1960s, newer and more radical organizations joined in the environmental crusade. Springing from a group of Americans and Britons protesting nuclear weapons testing, Greenpeace quickly grew into a powerful international environmental organization dedicated to preserving all life, especially sea mammals. Disgruntled Greenpeace members who opposed the organization's stated policy of nonviolence formed radical groups such as Earth First! and the Sea Shepherds Conservation Society to increase the awareness of their crusades through the use of violent, attention-grabbing tactics, such as ramming whaling ships (Nash 1989, pp. 179–180; Kuipers 1992).

In the late 1960s, media coverage of ecological disasters reached Americans nationwide. Television provided dramatic images of visible destruction and offered implicit warnings that although an environmental disaster might be miles away today, a similar tragedy might strike the viewer's neighborhood tomorrow. The 1967 Torrey Canyon oil spill in England was followed two years later by the blowout on an oil rig off the coast of Santa Barbara, California. Images of oil-strewn beaches outraged many Americans and escalated public concern that the government was not sufficiently attentive to environmental issues (McCormick 1989, pp. 57–58). In that same year, the pollution-ridden Cuyahoga River in Cleveland, Ohio, burst into flames. As one commentator wrote, the images of the five-story high flames "demonstrated to the people of Cleveland and the nation as no scientific study or news report ever could that the burden being placed on the environment was reaching limits that could be crossed only at the peril of the future" (Gordon 1993, p. 32).

Media coverage of other ecological disasters in the 1970s increased public awareness of a variety of potential environmental hazards. The apparent link between the Hooker Chemical and Plastics Company's former waste site and medical problems among residents of Love Canal, New York, awakened Americans to the potential dangers of hazardous waste (Paehlke 1989, p. 39).[17] Similarly, the accident at a nuclear power plant at Three Mile Island, Pennsylvania, in March 1979, created an uproar over the risks of atomic energy (Petulla 1980, p. 161).

Heavy television coverage of such events brought environmental issues into the homes of millions of Americans in the simplistic but powerful form of visual images. Environmental activism, academic publications, and educational programs could introduce ecological ideas to many people, but no organized attempt to stimulate public interest was more effective than the dramatic pictures of what appeared to be unanticipated ecological disasters.

2
Understanding Contemporary Environmentalism

THE MODERN ENVIRONMENTAL movement is a product of historical events of the past two centuries. But the vast social, economic, technological, and ideological changes of the 1960s and 1970s transformed the way Americans think and act about the environment. They also helped turn the relatively small-scale conservation movement into the immense and powerful environmental movement. The shifts in thought follow directly from the preservationist and conservationist roots of early environmentalism, yet the intensity and pervasiveness of the new movement has been unparalleled.

Although the modern environmental movement consists of many different factions, two prominent ideological views, ecocentrism and technocentrism, form the core traditions of environmental thought.[1] Ecocentrism follows from preservationism and the philosophical underpinnings of romantic transcendentalism. According to this ideology,[2] all organic material has intrinsic value and nature possesses its own purposes and rights. Rejecting anthropocentric worldviews, ecocentrists focus on the value of all nature, not just humans. Humans are only one species and deserve no special privileges or rights over others (Oelschlaeger 1991, pp. 294–296).

Roderick Nash describes the evolution of this approach in *The Rights of Nature* (1989). Nash traces the development of the scope of ethical inquiry from concern only with oneself to concern for one's nation and one's species. He sees great promise in the emerging field of environmental ethics, which enlarges notions of right and wrong and proposes the extension of rights beyond humankind to include all of nature.[3]

Often reflecting a sense of urgency, ecocentrists may rely on alarmist tactics to champion their cause. Because the predominant worldview remains anthropocentric, some ecocentrist groups such as Earth First! and the Sea Shepherd Society believe that they must take drastic measures to save the earth from human depradation.

An alternative perspective is technocentrism. With ideological roots in progressive conservationism, its guiding principles include rationality, progress, efficiency, and control. Technocentrism is unashamedly anthropocentric, advocating the position that humans have special rights and privileges over the rest of nature. Also known as the economic perspective, proponents of this school of thought often rely on cost-benefit analysis to measure the value of wilderness or to weigh the expected utility of a proposed environmental initiative. Technocentrists view wilderness as distinct from civilization and valuable only as determined by economic markets (Oelschlaeger 1991, pp. 286–288).

Drawing heavily from the doctrines of scarcity associated with Malthus, Ricardo, and others, some adherents of this utilitarian wing of environmentalism believe in the importance of government intervention to reach environmental goals that would be impossible to accomplish by unrestricted private markets (Petulla 1977, p. 36). Franklin D. Roosevelt championed this philosophy during the New Deal era. Creating the Civilian Conservation Corps and other large-scale environmental programs, FDR justified large-scale government intervention on behalf of the environment by exploiting public dismay over the consequences of the laissez-faire approach of the Hoover administration (Petulla 1977, p. 32).

Because environmentalism is a complex movement that encompasses disparate philosophies and ideological perspectives, environmental ideologies are best understood as a range of beliefs. The ecocentric and technocentric traditions represent the two extremes of thought on the wide continuum of ecological ideas.[4]

The Movements within the Movement

The term *environmental movement* means different things to different people. Some authors stress the role of environmentalism as an ideology (Paehlke 1989), others equate the movement with a religion (Coffman 1994), and still others define environmentalists as political and moral crusaders (Rubin 1994, 10–11). Joseph Petulla, on the other hand, proffers a more positive and general definition of environmentalism as "the ideas and activities of those concerned with the protection or proper use of the natural environment or natural resources" (Petulla 1980, x).

Because environmentalism is such a broad term, the environmental movement naturally includes a wide variety of organizations representing an abundance of perspectives and concerns, ranging from the mammoth World Wildlife Fund to organizations seeking protection of a single species, such as the Friends of the Sea Otter. Environmentalism, then, is really a compilation of many movements, some with competing goals.

Explaining some of the distinctions between various groups, Charles Snow classifies conservation and environmental organizations by their size, issue orientation, and level of organization.[5] The diversity of organizational structures is dramatic, ranging from recreational groups like local rifle clubs to immense national and international organizations such as Greenpeace and the Sierra Club.

Support for environmental groups surged in the 1980s, partly in reaction to Ronald Reagan's campaign promises to streamline big government and provide relief from government regulations. Reagan's controversial appointments of James Watt to be secretary of the interior and Anne Gorsuch Burford to administer the Environmental Protection Agency fueled this growth by stirring controversy and inciting activists to intensify their lobbying efforts (Landy, Roberts, and Thomas 1990, pp. 245–251; Stanfield 1985, p. 1350).

Membership statistics highlight the extent of the growth of environmental organizations in the 1980s. In 1982, about seventy-five national environmental groups were active in the movement (Douglas and Wildavsky 1982, p. 128). By the early 1990s, a journalist who had spent his career covering ecological issues estimated the number of national groups had expanded to 150 and that local organizations numbered at least twelve thousand (Sale 1993, p. 92). The 1994 edition of Gale's *Encyclopedia of Associations* (1993) listed 470 groups under subject headings that related directly to environmental issues; the total would increase if we added the many other organizations with environmental concerns, most notedly lobbying organizations for conservation groups.[6]

The dramatic rise in strength and power of national environmental organizations is best demonstrated by examining the growth of the "Group of Ten," a loose federation of the most influential environmental groups, whose leaders began meeting regularly in 1981 to share ideas and collaborate. According to data collected in 1985, the ten organizations had a combined membership of 5,774,000 members and total budgets of $113.5 million (Stanfield 1985, p. 1352). By 1994, their size and strength had grown to 8,534,000 members with combined budgets of $315 million dollars (*Encyclopedia* 1993).

Although large national and international organizations are the most visible environmental activists, local citizen groups have been very successful in achieving their policy goals. From organizers who began the furor over hazardous waste at Love Canal to local groups protesting waste disposal sites in their neighborhoods, grass-roots activists work to protect their communities' environment.

One example of a successful grass-roots campaign was the fight against the use of Styrofoam food containers by McDonald's restaurants. Concerned about the health hazards created by the incineration of Styrofoam, one-time Love Canal activist Lois Gibbs organized a campaign through her Citizen's Clearinghouse for Hazardous Wastes. McDonald's customers from around the country mailed their used Styrofoam containers to the company's headquarters. The campaign succeeded: McDonald's discontinued use of all Styrofoam products and switched to paper packaging (Dowie 1992, pp. 110–11).

Environmental activists employ many different tactics to sway public opinion and policymakers. One major tactic is the move toward litigation to fight policies, regulations, and private-sector operations that are believed to threaten environmental quality. The Environmental Defense Fund (EDF), established in 1967, was the first organization created specifically for environmental legal advocacy. Along with the similarly oriented National Resources Defense Council (NRDC), the EDF instigates many lawsuits and provides a public forum for adjudicating environmental issues (Mitchell 1991, pp. 88–90). Suing the government on behalf of concerned citizens and the environment, the EDF and NRDC capitalize on a powerful and effective form of political activism. Sutherland and Parker's research on the development of environmental law presents data that validate a litigator's seemingly hyperbolic claim: "in no other political and social movement has litigation played such an important and dominant role. Not even close" (1988, p. 181).

Critics argue that the movement has won many of its battles by relying on "junk" science in the courts (Huber 1991). In response, various conservative groups have pushed to raise the standards that now exist for the admission of "scientific" evidence in product liability cases. They have also fought for legislation that would enable the developers of various products to countersue where charges are deemed frivolous.

Another important organizational tactic of environmentalists is the use of crises (real or exaggerated) to attract public attention. Describing the United States as "an environmentally conscious but confused nation," Stephen Klaidman argues that "environmentalists have always felt forced to manufacture crises and exaggerate risks to provoke political action" (1991, p. 73). Although he distributes blame among environmental activists, the public, the media, and others, he demonstrates that the environmental movement moves forward by crisis-mongering.

Many "environmental crises" are indeed real threats to quality of life. Unfortunately, a number of high-profile environmental campaigns distort the facts and exaggerate the dangers to mobilize interest and donations. A *New York Times Magazine* article (Bonner 1993) tells an alarming story about the political maneuvering that led to a ban on the sale of ivory. As late as 1988, conservationists agreed that imposing a ban on the sale of ivory would not be a wise ecological policy. Only one year later, all major wildlife organizations officially advocated a total ivory ban. Why the change? Leaders of wildlife organizations discovered that the cause of endangered elephants brought money and members: "The crusade to ban ivory was marked by increasing hyperbole and gruesome pictures of mutilated elephants and by public emotion and politics. Groups that endorsed a ban found that there was money in elephants—big money—and, conversely, that failing to join the crusade might cost them members" (Bonner 1993, p. 17).

According to Bonner, leaders of prominent organizations, including the World Wildlife Fund, admit that they caved in to pressure to join the crusade for an ivory ban. Through direct-mail campaigns and print advertisements, conservation groups pleaded for the ivory ban on behalf of endangered elephants, and membership rolls and donations increased dramatically.

A number of African countries, including Zimbabwe, Botswana, and Namibia, opposed the ban from the beginning as too rigid in light of a growing human population and the destruction of agricultural land by elephants. They have been permitting limited hunting under the supervision of communities that have been granted the right to control such activity. Thus far it does seem that controlled hunting has been more successful than a total ban in striking an effective balance between preserving elephant populations and meeting the needs of African peasants, though the issue remains controversial (Pearl, 1995; Sugg 1996). In June 1997, however, the 138-member United Nations Convention on International Trade in Endangered Species (CITES) approved the application of Zimbabwe, Botswana, and Namibia to sell an annual quota from their ivory stockpile. The sale, to Japan, is regarded as an experiment and will be associated with the development of new rules to govern trade in ivory. The decision was a defeat for United States environmental groups (Duke 1997).

The ivory ban is only one example of inaccurate and exaggerated information fueling an environmental campaign. Citing research into spotted owl populations in California, Oregon, and Washington, Gregg Easterbrook describes the "illusion of impending owl extinction [as] a parable of modern environmentalism" (1994, p. 22). He demonstrates that spotted owls are not on the brink of extinction and have a much larger and more varied habitat than environmental groups claim. Similarly, the ban on dumping sewage into the oceans, asbestos removal, the landfill debate, the Alar apple scare, and the dioxin scare are products of exaggerated and often inaccurate public scare campaigns.[7] And, as we shall see, these are far from being the only ones.

Environmental Issue Cycles

Public concern for environmental issues reached record levels in the 1970s and 1980s. While there has been no retreat, there are some signs of increasing public skepticism about environmentalist claims (Ladd and Bowman 1995). The power of environmental activists, however, remains quite strong, and together with their supporters in the media they wield considerable clout in Congress as well as over public opinion. This power was amply demonstrated by the failure of the Republican Congress in 1995 to overturn or modify a significant number of environmental regulations. These initiatives roused many environmental groups and their supporters to action, and they successfully beat back the Republican assault.

What accounts for the prolonged interest in environmental issues? Will public attention to ecological issues increase, remain stable, or decline? The lack of attention to particular issues raises a third, related question: Why are some ecological issues more attractive to the public than others?

In his well-known essay on the political economy of the environment Anthony Downs (1972) discusses the dynamics of issue cycles. He cites two principle reasons for the upsurge of environmental concern that marked the beginning of the modern ecological movement. First, the benefits of technology produced genuine ecological damage, which was apparent to the general public. For instance, one does not need special scientific training to recognize smog contaminating the air or trash polluting the waterways. Second, Downs argues, environmental concern was increased by an outburst of "ridiculous and exaggerated rhetoric" surrounding ecological debates (Downs 1972, p. 10).

Using the environment as an example, Downs tests a model of an issue-attention cycle that follows public issues through five stages of increasing salience, climax of interest, and decline. The first is the pre-problem stage, when most people are unaware that a problem even exists. The second stage of "alarmed discovery and euphoric enthusiasm" is followed by a period of recognizing the cost of progress (third stage). As people begin to weigh the alternatives involved in finding solutions, a fourth stage of gradual decline in interest follows. Finally, during the post-problem stage, the issue is no longer publicly salient, yet public understanding of the problem is forever changed.

Although Downs's model failed to predict the intensity of sustained concern for environmental issues, his analysis, somewhat revised, can help explain the rising tide of environmentalism. Downs finds that environmental issues are well suited to stirring public opinion. Because various ecological problems have appeared that could threaten the well-being of almost everyone, these issues can move rapidly from the pre-problem stage to the period of discovery and enthusiasm. The causes of environmental degradation—at least on the level of public discourse—appear straightforward. Advocates can identify a set of perpetrators (usually industry) who serve as villains in a public morality play. The broad objectives of the movement allow new issues to enter the cycle. Finally, the onset of environmental regulations and the threat of sanctions create an entire industry devoted to environmental cleanup that develops a vested interest in perpetuating environmental crises and consequent regulation. Of course, as Ladd and Bowman (1995) have pointed out, a general expression of support for environmental activism does not tell us how much the public would actually be willing to sacrifice as an economic tradeoff. Thus far the public has assumed that it can have its environmental cake and eat it too. There are signs, however, that this belief may be changing.

Issues of Declining Prominence

A small category of environmental issues have received less public attention recently than they did in the early years of the modern environmental movement. Activists and organizations have limited resources, so they must prioritize the problems they choose to address. Because of recent historical developments and the passage of legislative programs mandating clean air and water, among other things, the socialism-capitalism debate and the issue of nuclear power have declined in significance compared to other environmental concerns.

SOCIALISM-CAPITALISM

Many environmental activists in the 1960s and 1970s criticized the effects of capitalist economies on the environment. Popular textbooks such as Paul Samuelson's *Economics* introduced an entire generation to the belief that capitalism threatened environmental goals; the writings of Barry Commoner and other ecologists linked pollution to profit motives and advocated further state control and ownership to protect the environment.[8]

The historical record provides reasonably clear evidence that state control of industry does not necessarily guarantee environmental protection. If anything, evidence from the former Soviet Union and Eastern Europe demonstrates that Communist regimes inhibited ecological initiatives. With the breakup of the Soviet bloc, the international community is beginning to learn the extent of Soviet and Eastern European environmental deterioration. While under Communist rule, these countries' environmental quality declined rapidly, and government planning generally ignored or hindered environmental regulation of industry. Because the regimes quashed public protest and limited information flow, the international community and even most members of these societies were unaware of the extent of unchecked ecological hazards.[9] Murray Feshbach and Alfred Friendly summarize the extent of Soviet ecological destruction, what they call "ecocide," with these daunting words: "No other great industrial civilization so systematically and so long poisoned its land, air, water, and people. None so loudly proclaiming its efforts to improve public health and protect nature so degraded both" (1992, p. 1).[10]

Some commentators have noted a link between increased environmental activism and the overthrow of Communist regimes. In some of these countries, various nationalist movements linked together to support heightened environmental protection, protesting the ineffectiveness of Communist and socialist regimes to address ecological problems. Fred Pearce describes the power of recent

Eastern European environmental activism, naming ecological protest as a contributing factor in the overthrow of Communist regimes in Poland, Hungary, Bulgaria, and other former Soviet-bloc nations (Pearce 1991, pp. 107–131). Charles Ziegler chronicles the rise of environmental activism in the final years of the Soviet Union and finds similar political dynamics. Discussing the interaction between nationalism and ecological concerns, Ziegler notes the importance of environmental issues, particularly campaigns against pollution, in political protests in Azerbaidzhan, the Baltic states, Moldavia, and other regions (Ziegler 1987, pp. 117–125).

The irony of some American environmentalists' claims that socialism would promote their goals is that the very system they supported stifled citizen social movements similar to their own:[11] "The massive failure of authoritarian socialist countries to address the problem of environmental degradation has been convincingly demonstrated. These highly centralized and bureaucratized systems lack the flexibility and responsiveness to deal effectively with complex issues. . . . Responsible figures in the Soviet artistic and scientific communities resisted ecologically disastrous practices, but lacking the political freedom to mobilize public opinion against official policy, their cautionary voices could be easily ignored" (Ziegler 1987, 128–129). Needless to say, the argument that capitalism is responsible for pollution is not heard much these days, though businessmen and their selfishness continue to be berated.

NUCLEAR POWER

The antinuclear campaign began with protest against atomic weapons and later grew to include activism against commercial nuclear power as well. The Atomic Energy Commission (AEC), the government commission established by the McMahon Act to regulate the nuclear industry, began releasing nuclear testing information in the mid-1950s. In response, small but vocal groups began actively protesting above-ground nuclear weapons testing, combining antimilitarism with concern for the environment. Studying the estimated public health effects of radiation, scientists such as Barry Commoner were first introduced to ecological issues through this movement (Novick 1969, pp. 197–201).

Opponents of nuclear testing won their battle. Scientific evidence in the 1950s demonstrated the serious health hazard of fallout from atomic tests. In response, the United States, the Soviet Union, and Great Britain agreed to a test ban that lasted from 1958 until 1961. After two years of renewed testing, the three countries signed another treaty, the 1963 Atmospheric Test Ban Treaty, which is still in effect today (Eisenbud 1973, p. 56).

With the eradication of atmospheric atomic testing, the locus of protest moved from weapons testing to nuclear power.[12] In the 1960s, the concern about atomic energy escalated. Arthur Tamplin and John Gofman, scientists at the AEC, resigned their positions when the commission pressured Tamplin to suppress some of his findings (Jasper 1990, pp. 110–111). The two scientists became active in the antinuclear movement, publishing evidence of the deleterious effects of low-level radiation in their influential book *Poisoned Power* (Gofman and Tamplin 1971). As the controversy over the safety of atomic energy escalated, national organizations such as Friends of the Earth entered the debate, bringing further public attention to what they perceived as the potential global environmental hazards of nuclear power.

In the aftermath of the 1979 accident at Three Mile Island, the nuclear power industry went into a downward spiral (though the decline had actually begun in the early 1970s). Although no one was injured in the accident (indeed, investigators concluded that fallout from the accident could, at most, have resulted in only one additional cancer death), Three Mile Island received more publicity than earlier accidents. Rothman and Lichter's (1987) study of risk assessment and nuclear energy found that hostile media reporting of atomic energy, which failed to reflect the attitude of the majority of scientists regarding nuclear power, continued into the 1980s.

Although the nuclear power controversy remains a point of tension between environmental activists and scientists, the issue has declined in salience over the past decade. At the height of the development of atomic energy in the United States, companies ordered 153 plants between 1970 and 1978. In the years between 1974 and 1984, however, 103 orders were canceled (Scheffer 1991, p. 79). An unofficial moratorium on new nuclear plants continues even today: "No U.S. utility has ordered a reactor since 1978. Why fight a battle in which any self-professed environmental group—or just a few people scared of nukes in their community—can tie up a project indefinitely merely by hiring a lawyer?" (Faltermayer 1988, p. 105).

In fact, a substantial majority of energy experts still believe that nuclear power can play a useful role in an energy mix and do not believe (as we shall later see) that nuclear plants pose a threat to public health. And, of course, a number of countries, including France, have continued to rely on nuclear energy, which is produced safely and relatively inexpensively.[13] Today, most of the controversy involving nuclear energy focuses on the disposal of nuclear waste. Such is the anxiety with which radioactivity has come to be regarded by the public that it has become almost impossible to find repositories for both high-level and low-level nuclear waste. The latter issue is threatening even to restrict the use of radioactive materials in medical research and treatment (Burns 1988; Mahlin 1995).

Other Current Environmental Issues

Many environmental issues have remained prominent throughout the past quarter-century of contemporary environmentalism. Despite the slow beginnings of federal action, Congress and successive presidents have addressed some of the most serious environmental issues facing Americans. In each of these policy areas, Congress has enacted comprehensive legislation intended to confront important ecological problems.

The National Environmental Policy Act of 1969 (NEPA) was a pioneering legislative effort that marked the beginning of a strong federal role in environmental regulation. Among other things, NEPA required the completion of environmental impact statements, studies that estimate the ecological consequences of a proposed project, for all new federal enterprises. NEPA also created the Council on Environmental Quality, a board of advisers to inform the president and Congress about environmental issues. In the wake of these developments, President Richard Nixon established the Environmental Protection Agency (EPA) in 1970 as an independent agency that reports directly to the president and oversees regulations pertaining to environmental quality (Whitaker 1976, pp. 48–56). These developments signaled the beginning of a new era of federal environmental protection.

POLLUTION CONTROL

In the area of pollution control, federal actions moved from voluntary recommendations to direct regulation. Federal assistance with sewage treatment began in 1948, and Congress passed a few air pollution bills in the 1950s, but serious efforts to combat pollution did not begin until the late 1960s (Vig and Kraft 1984, p. 9).

Of the two major forms of pollution legislation, clean-air issues have received more public attention than water pollution. Although automobiles were a recognized source of air pollution as early as the 1940s (Whitaker 1976, p. 95), the first national legislation directly addressing car emissions, the Motor Vehicle Air Pollution Control Act, did not pass until 1965. This legislation created federal auto emissions standards and marked the beginning of specific national standards of pollution regulation. The landmark Clean Air Act of 1970, which Congress passed with near-unanimous votes in both houses,[14] established ambient-based emissions standards that emphasized the need for environmental regulations to protect public health. Seven years later, Congress adopted important amendments to the act, described as "one of the most detailed environmental laws ever written" (Tobin 1984, 228). These provisions gave the EPA power to define acceptable levels of air pollution and fine excessive polluters to reduce the profit motive of noncompliance, and gave states more flexibility in implementing the act.[15]

Despite Reagan administration attempts to weaken the Clean Air Act, public support for stringent air pollution controls helped convince legislators not to relax air quality standards. Contrary to the administration's beliefs, Ronald Reagan's electoral victory in 1980 was not a mandate to deregulate the environment. Public pressure called for stronger, not weaker, antipollution laws: "At midterm the *Wall Street Journal* noted that the political climate had changed so much that the administration clearly was on the defensive and willing to delay attempts at change in order to avoid the adoption of amendments that would strengthen the law. . . . Harris poll results confirmed that a plurality (38 percent) of Americans were pleased with existing levels of air pollution regulations, and three times more Americans favored strengthening the laws than favored relaxing regulations" (Tobin 1984, pp. 231–233).

George Bush, Reagan's successor, labeling himself the "environmental president," supported further expansion of the Clean Air Act. In the first significant revisions for over a decade, the Clean Air Act of 1990 passed with wide bipartisan support. The act provided comprehensive new measures for combating pollution, including provisions that reduced smog, tightened automobile emissions standards, limited emissions of sulfur dioxide and nitrogen oxides, gave the EPA power to set emissions limits on toxic chemicals, banned CFCs by the year 2000, and provided benefits to workers displaced by environmental initiatives (Weisskopf 1990b).

Water pollution regulation has not been as high-profile an issue over the years. When Congress delayed action on clean-water legislation in the early 1970s, President Nixon signed an executive order requiring government permits for industrial pollution of navigable waters (Whitaker 1976, pp. 79–80). The Clean Water Act of 1972 (CWA) created technology-based standards for pollution levels, funded the improvement of waste-water treatment facilities, required discharge permits, and established a mechanism for land-planning programs to decrease nonpoint runoffs such as from city streets or farms. Although the CWA was amended slightly in 1977, Congress gave the Clean Air Act priority over water-pollution regulation.[16]

Federal policies to combat air and water pollution have been a qualified success. Between 1970 and 1980, air pollution decreased by 21 percent and the number of days with reported "unhealthful" air quality declined by one-third (Vig and Kraft 1984, p. 17).[17] The decline continued through the 1980s and into the 1990s. At that point, however, new controversies emerged as the EPA moved to initiate new rules primarily concerned with the dangers of newly examined smaller particles. Again, battle lines emerged both because of disagreements among scientists over the level of danger these particles presented and because of differing estimates of costs. Although the EPA had been given the authority by Congress to renew and amend regulations, many businessmen and congressional Republicans, as well as a number of Democrats from the industrial Midwest, such as John Dingell of Michigan, argued that the agency was arrogating too much authority to itself.[18] Nevertheless, in June 1997 President Clinton, accepting the arguments of Vice President

Al Gore and Carol Browner, the new chief of the Environmental Protection Agency, approved significantly tighter standards on air pollutants.

From 1970 to 1983 American water quality did not decline despite rapid economic growth and expansion during the decade (Ingram and Mann 1984, pp. 252–53). In general, the period of the 1980s and 1990s have been associated with at least some—and often dramatic—improvement in water quality. Nevertheless, taking advantage of a hitherto largely ignored provision of the 1972 Clean Water Act, environmental groups have been winning cases in federal courts designed to further tighten water pollution controls (Cushman 1998). In collaboration with the Clinton administration and the EPA, efforts are being directed at more effectively controlling runoffs of nitrogen and phosphorus. Care is also being taken to better control runoffs from large corporate farming conglomerates, and to expand wetlands.

Although these statistics demonstrate that federal regulations have positively affected the pollution problem, most environmentalists are not satisfied with current policy efforts. Passing environmental regulations, they argue, is only a first step in combating pollutants; administrations must also actively enforce regulations. For example, environmental organizations originally praised the passage of the Clean Air Act of 1990 as a great success, but the same groups later criticized the Bush administration for relaxing regulatory efforts during economic downturns. Additionally, many environmental laws contain regulatory limits that are beyond the capability of modern technology. In such instances, Congress usually extends compliance deadlines, and frustrated environmentalists then criticize the lawmakers for caving in to industry pressure.

The most radical activists deride all policies of pollution control, arguing that regulations provide a government license to pollute. Instead, these environmentalists call for complete bans (along the order of the DDT ban) to eliminate hazardous pollutants. To those who advocate pollution prevention as the only acceptable form of pollution control, government pollution regulations will always represent policy failures. On the other hand, criticism of what are perceived of as excessive expenditures and bureaucratic waste involved in the enforcement of current laws has also increased in recent years, especially since the 1994 elections (Easterbrook 1995).

Because of the development and widespread use of synthetic chemicals, Congress passed the Federal Insecticide, Fungicide, and Rodenticide Act of 1947 (FIFRA). This act originally emphasized consumer protection but failed to provide regulations for pesticide use.

In the wake of the controversy generated by *Silent Spring* (Carson 1962), Congress amended FIFRA in 1964 to direct government attention to the potential deleterious public health effects of excessive pesticide use. After further study and active lobbying by environmentalist groups, FIFRA was again amended in 1972 to specify criteria for eliminating the use of substances that pose "any unreasonable risk to man or the environment, taking into account the economic, social, and environmental costs and benefits of the use of any pesticide" (quoted in Moore 1987, p. 17). FIFRA empowered the EPA to ban or restrict any substance that the agency determines poses a significant health or environmental risk (Schroeder 1991, p. 269). Subsequent legislation dealt with the scientific methods of determining the potential dangers of chemicals (Moore 1987, p. 19).

Under current law, chemical companies must extensively test prospective pesticides before receiving EPA approval. In a series of experiments ranging from initial screening for toxicity to long-term animal toxicology testing and experimental field studies, manufacturers typically spend millions of dollars and test products for years before the EPA will even consider approving a new pesticide (Whelan 1992, pp. 151–152). The end result of meeting government regulations ensures that approved products meet "tolerance levels for pesticides that allow a safety buffer of hundreds (sometimes thousands) of times of dosage thought to be harmful" (Jukes 1993, p. 7).

Although a few radical activists would define success only as a complete ban of all synthetic chemicals, the environmental crusade against pesticides represents one of the greatest policy successes of the modern movement. Over the course of a few decades, the federal government has successfully imposed and implemented extensive regulations to control pesticide use and has banned the use of many pesticides.

Despite the extent of current federal pesticide regulations, some environmentalists continue to call for further regulation of pesticides and synthetic chemicals. On the other hand, a number of respected scientists maintain that the dangers of pesticides have been highly exaggerated.[19] The most serious arguments revolved about the use of the Delaney clause as the basis for pesticide regulation. The EPA's interpretation of the clause, which required the banning of any substance suspected of being carcinogenic, involved the administration of maximum dosages of suspected carcinogens to rats.[20] If the result of such administration was an increased number of tumors at any site in the rat's body (even if other kinds of tumors decrease in number) the proposed substance was considered dangerous to human beings.

A variety of attacks were leveled against this procedure. Authors such as Elizabeth Whelan (1992), Edith Efron (1984), and Aaron Wildavsky (1995) argued that government regulations relied on inaccurate or misrepresented scientific data that overestimated the risk of potentially toxic or carcinogenic substances. No one, however, has written more on this subject than biochemist Bruce Ames, the inventor of the Ames test. Ames argued that under the standards created by the EPA pursuant to the Delaney clause, literally thousands of natural substances would be considered dangerous carcinogens (Ames, Profet, and Gold 1990a, 1990b; Ames and Gold 1990, Ames 1991; Ames 1992b). The National Research Council recommended that the Delaney-clause standard be replaced by the concept of "negligible" risk, an assessment supported by the EPA, the Food and Drug Administration, and the Department of Agriculture (Wildavsky 1995; Easterbrook 1995). In 1995 the Republican-controlled 104th Congress introduced legislation to ban the use of the Delaney clause as the standard for judging carcinogenicity. A 1996 NAS report argued strongly that the dangers from carcinogens in food were not substantial (National Research Council 1996). In the same year legislation was passed setting up a standard of "reasonable certainty of no harm," and the EPA consequently announced a decision to move away from animal testing and to rely more on microbiological techniques (Cushman 1996a). It is too early to determine whether or not this standard will lighten the burden imposed by the Delaney clause. It may even raise it (Tolman 1996).

In fact it already has. In 1993, the National Research Council had pointed out that there were few data on the pesticide toxicity and exposure levels relevant to children and called for further research into the matter. Congress rushed to pass legislation calling for a tenfold margin of safety beyond what would normally be the case for such protection. Despite the 1996 National Research Council report mentioned earlier and a more recent report by the American Cancer Society, which also does not call for special measures and which argues that dangers from pesticides to human health are all but negligible, pressure is rising for new and more expensive tests and rigid standards (*New York Times,* November 16, 1997; Cushman 1997).

HAZARDOUS WASTE

One of the consequences of technological and economic process is the creation of massive quantities of waste, some of which is hazardous and all of which creates disposal problems.

The first comprehensive legislative act to address the issue of hazardous waste, the Resource Conservation and Recovery Act of 1976 (RCRA), created regulations to control the disposal of hazardous waste but made no effort to eliminate it. RCRA's provisions gave the EPA power to identify hazardous wastes, created a registration system to track waste from the point of its generation through its disposal, regulated its transport, established dumping standards, and mandated state involvement in the process.

Publicity surrounding chemical dumping at Love Canal created widespread national concern about the dangers of hazardous waste and liability for cleaning up dangerous sites. In response, Congress enacted CERCLA, the Comprehensive Environmental Response, Compensation, and Liability Act of 1980, popularly known as Superfund. Providing federal guidelines for the cleanup of hazardous waste sites, Superfund established parameters to determine liability for cleanup and damages.

Additionally, Superfund enabled the federal government to pay cleanup costs when the source of the pollution was undetermined or the company responsible for damages was no longer solvent. The Superfund legislation also called for the creation of more waste disposal sites, but public fears of the health effects of living near such sites inhibited EPA progress. In large part because of local protests against new waste sites, the implementation of Superfund has failed to create sufficient disposal space for America's hazardous waste.

By almost all accounts, hazardous-waste legislation has been a failure. Many environmentalists criticize the programs for doing too little to discourage the generation of hazardous wastes. Additionally, they complain about the slow pace of cleanups and the lack of adequate funding appropriated for them. Although the EPA has spent over $11 billion on Superfund cleanups during the past decade, only sixty sites are officially clean. The EPA has provided emergency (but not full-scale) cleaning at four hundred sites. Another twelve hundred are still untouched, with thousands of potential sites still to be added to the Superfund list (Passell 1991b). Other critics argue that cleanup standards are too rigorous, that demands for soil to be left "clean enough to eat" are not realistic, and that vast outlays required for this could be spent more effectively elsewhere (Landy and Hague 1992; Easterbrook 1995; Wildavsky 1995; Probst et al. 1995).

Various controversies surrounding Superfund, including corporate liability and proposed risk tradeoff provisions, have thus far prevented renewal of the legislation. Though the Republicans have promised that new legislation will be brought to the floor in the 105th Congress (1997–98), their attempts have been plagued by the same controversies (Freedman 1997).

ENDANGERED SPECIES

Estimates of the number of species on earth vary widely from three million to thirty million (Wilson 1991, p. 4). Just as we cannot closely approximate the number of species, so we can only estimate the rate of species extinction. Unlike many environmental problems, however, species loss is irreversible. No amount of regulation or government spending can bring back an extinct plant or animal, though species with very few remaining members (the American bison, the sea otter, the elephant seal, etc.) have quickly increased in number after appropriate action.

Advocates of strong species-protection measures lament every species casualty as a lost opportunity that forever changes the balance of nature. Newly discovered or utilized species may have a variety of beneficial uses as foods, medicines, fibers, or other products. The paradigmatic case environmentalists use to illustrate the importance of exotic species is the rosy periwinkle. Researchers discovered that this small flower is the source of medicines used to treat Hodgkin's disease and lymphocytic leukemia. Although environmentalists may argue that the plant and animal species that become extinct today might offer hope for a pharmaceutical breakthrough tomorrow, the case of the rosy periwinkle is not necessarily convincing. Originally found only in Madagascar, the rosy periwinkle is actually easy to cultivate and is a common garden plant that now grows wild throughout the warmer regions of the world, including the southern United States (Mittelbach and Crewdson 1994). Consequently it has never been considered threatened or endangered. Of course those who believe that animals and plants have a right to survival see no need to justify the saving of species by pointing to projected human advantages.

The federal government has addressed the species issue in two ways: habitat protection and government regulation. National parks and wildlife refuges provide a level of protection for plants and animals living within the borders of a reserve area. Federal environmental regulations extend protection to endangered species on public and private lands.

The first federal legislation specifically targeting endangered species, the Endangered Species Preservation Act of 1966, provided funding for global species preservation. Three years later, this law was amended to criminalize the killing of animals whose species are known to be in danger of worldwide extinction (Caldwell, Hayes, and MacWhirter 1976, pp. 143–145). It was not until 1973, however, that Congress passed comprehensive species protection legislation. The Endangered Species Act of 1973 (ESA) helped fund state conservation programs, pledged American cooperation with other countries on behalf of endangered species, and empowered federal agencies to maintain lists of officially threatened species and to block any land development or other land usage that threatens habitat loss (Williams and Nowak 1986, pp. 131–132).

The Endangered Species Act not only gives the federal government significant power to act on behalf of threatened species but also provides grounds for environmental organizations to sue potential violators and initiate legal action to stop developments that may threaten endangered species. In the two decades since ESA was passed, environmental organizations have used this power to shape the act's judicial implementation. Court rulings redefined the application of the act to such a degree that even Senator Mark Hatfield, a co-author of the 1973 legislation, recognizes that "[t]here is no question that the act is being applied in a manner far beyond what any of us envisioned when we wrote it twenty years ago. . . The fact is that Congress always considered the human element as central to the success of the ESA. . . . The situation has gotten out of control" (quoted in Coffman 1994, p. 128).

For many environmentalists, the evolution of federal efforts to protect species is clearly a success. The National Research Council's report on ESA in 1995 also gave it high marks (*Science* 1995a, p. 1124). On the other hand, Charles Mann and Mark Plummer argue that the act is too rigid. They contend that there is no reason to preserve all species and to legislate possibly unlimited expenditures to that end (Mann and Plummer 1995). After all, no more than 5 percent of the species that have appeared on earth have survived to the present day, and most of the rest died out before the appearance of mankind, though the rate of species destruction in the twentieth century is probably more rapid than at any other time in the history of the earth since the extinction of the dinosaurs. The number of species on this planet is not a given which cannot be disturbed for any reason. At this point in history, the decision to save a species is as much an interference with nature as is the failure to do so.[21]

The most recent development in the fight for endangered species is the international concern about biodiversity. President Bush made headlines in 1992 when he refused to sign the Convention on Biological Diversity at the Earth Summit in Rio de Janeiro, Brazil. Despite the lobbying efforts of environmental groups and the exertion of media pressure during the reelection campaign, Bush refused to sign the treaty, arguing that it violated the intellectual property rights of biotechnological research and development firms. His position received support from a number of environmental lawyers, who wrote in *Science* that "the treaty might just as appropriately have been designated the 'Convention on Biotechnology Transfer.'" They elaborated: "Major portions of the treaty, and certainly many of its key provisions, mandate that the signatory nations facilitate the transfer of technology among themselves and, particularly, from developed nations to less developed nations" (Burk, Barovsky, and Munroy 1993, p. 1900). Environmentalists finally achieved success when President Clinton signed the treaty in 1993.

GLOBAL WARMING AND STRATOSPHERIC OZONE DEPLETION

Two emerging issues, global warming and ozone depletion, concern the environmental impact of changes in the earth's atmosphere.

Scientists vehemently disagree over issues of global climate change. James Hansen, who popularized concern about global warming in the late 1980s, generated his hypothesis that the earth is warming after collecting data from surface thermometer readings over thirty-five years. Other scientists who measure the earth's temperature from satellite data find no definite warming trend (Rensberger 1993, p. 38). Clearly, models of global temperature change are incredibly complex and difficult to construct with precision; for example, scientists have to determine how much external factors—such as volcanic eruptions, ocean currents, and solar activity (*Economist* 1998)—affect weather patterns as compared to concentrations of greenhouse gases. Thus we have two issues. Is the earth warming and to what extent is this warming (if occurring) the result of human activity, primarily due to the growth in the production of carbon dioxide and other gases that trap heat in the atmosphere (greenhouse gases)?

Given the uncertainty of measurement techniques, the federal government has been hesitant to implement aggressive policies designed to combat global warming. Broad legislative programs such as the Clean Air Act include some provisions for controlling the discharge of greenhouse gases, but most government policies have emphasized the need for further study of global warming before implementing drastic policy changes. However, scientists' belief in the reality of a warming trend is growing stronger, though the importance of the role played by human activities is somewhat less certain

(Kerr 1995; Masood 1995; Murray 1996a; Stevens 1996; Michaels 1997; Houghton 1997; Burroughs 1997) and a powerful contingent of scientists still harbors strong doubts. Nevertheless, pressure for action is growing, despite President Clinton's deferment of curbs on greenhouse gas emissions at the June 1997 United Nations conference (*New York Times,* June 27, 1997, pp. 1ff).

Ozone depletion has also received increased public attention since the late 1980s, but is not as controversial an issue as global warming has been. Most scientists agree that the ozone layer over Antarctica has decreased in density in recent years. Because the stratospheric ozone is a protective layer of gases that screens much of the damaging radiation from the sun, ozone depletion has many potentially harmful effects ranging from increased incidence of skin cancer to agricultural damage.

The Clean Air Act of 1990 mandated the elimination of chlorofluorocarbons (CFCs), a common refrigerant that is implicated in the depletion of the ozone layer, by the turn of the century and gave the president the power to expedite the CFC ban.[22] The law actually implemented and even accelerated the Montreal Protocol schedule agreed upon in 1987.

After its passage, new scientific studies reported greater threats to the ozone layer than had been estimated earlier. "The new evidence [showed] not only a huge vortex of cold air favorable to formation of an ozone hole but the highest levels of ozone-damaging chlorine and bromine compounds ever recorded in the stratosphere" (Abramson 1992). Pressured by a unanimous Senate resolution, President Bush issued an executive order in 1992 to eliminate domestic production of CFCs by 1995.[23]

ACID PRECIPITATION

Congress acts swiftly to address some environmental issues, but legislators delay action on other potential ecological hazards, such as acid precipitation resulting from the emission of sulfur and nitrogen oxides into the atmosphere, which are transformed into sulfuric and nitric acid in rain water. The concern about acid rain is not new. References to potentially detrimental acid precipitation already appeared in the nineteenth century (Ray 1990, p. 50), but legislators are only now beginning to address the problem of acid rain in significant environmental initiatives.

Some scientists have argued for immediate government action to reduce emissions before acidification destroys aquatic life in thousands of lakes and threatens entire forests. But others cited existing evidence to show that acid rain is not a significant enough threat to justify massive government intervention. Faced with competing expert accounts and few clear policy options, Congress chose a middle ground by creating a federal research program in 1980 to study the problem. When the National Acid Precipitation Program's staff met for a final discussion of the program's report, the director summarized the results of the study, concluding that "[t]he sky is not falling, but there is a problem that needs addressing." Despite findings linking acid rain and ecological damage, the study reveals that "the amount of damage is less than we once thought, and it's much less than some of the characterizations we sometimes hear" (James R. Mahoney, quoted in Stevens 1990b, p. C1).

Congress finally implemented specific acid rain policies in provisions of the Clean Air Act of 1990, which established a pollution credit trading system. Under this policy, utilities receive a fixed number of allowances that permit the emission of sulfur dioxide, a chemical often linked with acid rain. Unused allowances can be traded at auction, with the proceeds used to pay for cleanup programs. Although proponents argue that permits provide economic incentives to reduce dangerous emissions while raising funds to defray costs, critics call them government licenses to pollute which do little to solve long-term problems. For some years the data was not sufficient to evaluate the effectiveness of the new program (Parrish 1993).[24] By 1996, however, even the director of the EPA's acid rain division praised the "overcompliance" that had resulted from the program (quoted in Bukro 1996).

SUSTAINABLE DEVELOPMENT

Some environmentalists have proposed policies of limited or even zero economic growth to lessen environmental destruction. Representatives of less developed countries, however, argue that such proposals unjustly discriminate against developing nations. Although industrialized nations have already benefited from rapid technological and economic expansion uninhibited by strict environmental regulation, they complain, the international community expects developing nations to bear the costs of environmental cleanup even as they try to modernize their economies.

Sustainable development is a reaction against no-growth ecologists. This doctrine aims to develop a relationship between economy and ecology that fosters controlled economic growth without unnecessarily increasing ecological damage. Indeed, proponents of sustainable development believe that policies to stimulate growth and development can actually promote care for the environment when planned accordingly.

Although proposals vary, many proponents call for governments to move beyond policies of environmental protection and work toward international cooperative efforts to combat the root causes of ecological problems. James Speth, president of the World Resources Institute, describes five transitions that he believes are necessary components of successful sustainable development: a demographic transition to stabilize population; a technological transition to move toward environmentally sound industries and agriculture; a social transition to provide more equitable distribution of wealth; a transition of consciousness to deepen appreciation and understanding of nature; and an institutional transition to facilitate international cooperation on environmental initiatives (Speth 1992, pp. 209–213).[25]

Into the Future: Emerging Ideological Perspectives

In the past decade, new ideological perspectives on the environment emerged, as other previously ignored ecological worldviews gained increasing acceptance.

ECOFEMINISM

One such emerging perspective, ecofeminism, applies feminist critiques to environmental practices. Spokespersons for this movement, most notedly Karen Warren and Ynestra King, have detailed ecofeminist theory in journals of ethics, philosophy, and women's studies.[26] Radical ecofeminists emphasize the differences between men and women, calling on everyone to embrace female relational and empathetic perspectives as the source for ecological healing. Because many ecofeminists believe that women are naturally closer to and more aware of nature, they advocate the need for "a period of matriarchy to redress the horrors of the patriarchal millennium" (Salleh 1992, 203).[27] Environmental degradation is best understood as yet another negative consequence of male domination; just as men seek to control women, so do they dominate and subjugate nature. As Karen Warren puts it, ecological feminism "provides a distinctive framework both for reconceiving feminism and for developing an environmental ethic which takes seriously connections between the domination of women and the domination of nature" (1990, p. 126).

Although the most radical ecofeminists warn against any participation in the male-dominated social order, more moderate adherents contend that women will only succeed in transforming flawed social institutions by working from within them (Lewis 1992, pp. 35–36). Seeking peaceful means to transform rather than reform society, ecofeminists see themselves as providing an alternative to Western patriarchy.

DEEP ECOLOGY/RADICAL ENVIRONMENTALISM

Another perspective of growing influence is that of deep ecology. A phrase originally introduced by Arne Naess in 1974, deep ecology rejects anthropocentrism and calls for an ecocentric understanding of the earth. It has close affinities with radical ecofeminism. Deep ecologists maintain that all beings are equally valuable. Egalitarianism should replace utilitarianism, for human-centered policies and programs fail to respect the intrinsic worth of all of nature (Sessions 1991). In contrast to the "shallow" ecology of mainstream environmental groups concerned only with surface issues such as pollution and conservation, deep ecologists believe that their perspective provides a new framework for understanding the natural world.

Vocal proponents of curbing population growth, deep ecologists advocate the return to pre-Industrial Revolution population levels. Additionally, they call for revolutionary systemic changes to create new socio-political institutions that advance ecocentric policies and perspectives (Nash 1989, p. 148).

On most policy issues, deep ecologists seek drastic measures to end detrimental human impacts on the environment. For example, most regard total bans as the only acceptable solutions for controlling environmental pollutants; such regulations as pollution limits and emissions controls allow destructive practices to continue and therefore implicitly endorse ecological degradation (Borrelli 1988, p. 74).

Although some committed environmentalists embrace deep ecology, it is "less a movement than a bundle of ideas held to a greater or lesser degree by a heterogeneous grouping of organizations and individuals" (Borelli 1988, p. 74). Deep ecology thus provides the ideological foundation for a new and rapidly growing wing of the ecological movement, radical environmentalism.[28] Disturbed by the rapid destruction of the earth and the anthropocentric goals of social and political institutions (including, they would claim, the mainstream environmental movement), radical environmentalists call for a complete restructuring of society. Martin Lewis explains the fundamental tenets of this perspective:

> The dominant version of radical environmentalism rests on four essential postulates: that "primal" (or "primitive") peoples exemplify how we can live in harmony with nature (and with each other); that thoroughgoing decentralization, leading to total autarky, is necessary for ecological and social health; that technological advance, if not scientific progress itself, is inherently harmful and dehumanizing; and that the capitalist market system is inescapably destructive and wasteful. These views, in turn, derive support from an underlying belief that economic growth is by definition unsustainable, based on a denial of the resource limitations of a finite globe. (Lewis 1992, p. 3).[29]

Although most radical environmental groups are young organizations, their memberships are expanding at dramatic rates. Greenpeace, generally considered the first and most conservative of the radical groups, is "the fastest-growing environmental activist organization in the world" (Lehman 1990, p. 2). Indeed, during the 1980s three radical groups—Greenpeace, the Sea Shepherds Conservation Society, and Earth First!—enlarged their memberships at a rate five times greater than that of nonradical groups (Mitchener 1991, p. 574). The number of donors to Greenpeace dropped off slightly in the early 1990s. Still, in 1995 their annual income was about 130 million dollars a year (*Economist*, 1995).

GAIA HYPOTHESIS

The Gaia hypothesis, formulated by the scientist and former medical school professor James Lovelock, takes its name from Gaia, the Greek goddess of the earth. The hypothesis, which fits within the deep-ecology framework, views the earth as a living organism; the planet and all things on it are part of an organic whole. Describing his hypothesis, Lovelock writes, "the atmosphere, the oceans, the climate, and the crust of the Earth are regulated at a state comfortable for life because of the behavior of living organisms. . . . The conditions are only constant in the short term and evolve in synchrony with the changing needs of the biota as it evolves" (Lovelock 1990, p. 9).

Lovelock argues that Gaia is both a scientific and a religious way of looking at the world that gives a new perspective on life (1990, pp. 205–207). Advocating a move away from urbanization and the return to a more natural state, Lovelock warns that human actions that destroy the earth may lead to their own destruction: "Gaia is not purposefully anti-human, but so long as we continue to change the global environment against her preferences, we encourage our replacement with a more environmentally seemly species" (Lovelock 1990, p. 236).

At first glance, Gaia appears more of a philosophy or religion than a serious scientific theory. One of Lovelock's recent books, designed to articulate and provide scientific evidence in support of Gaia, was published as part of Bantam's New Age series and echoed many spiritual themes.

Despite the philosophical and religious undertones of his work, the Gaia hypothesis is beginning to gain respect among some scientists, particularly atmospheric scientists (Coffman 1994, p. 43). Lovelock, a highly regarded scientist who has served as a consultant to NASA and other agencies in the United States and Britain, invented the electron capture device, an essential research tool that allows scientists to detect minuscule traces of substances to the parts per trillion (Rubin 1994, p. 240). His writings discuss Gaia without polemics, and he is willing to adjust his theory when subjected to criticism. In an ever-changing world, Lovelock's theory does provide a more plausible explanation of natural phenomena than the steady-state hypothesis, which views the earth as having fixed parameters that always work toward the restoration of the status quo ante.[30] Nevertheless, the scientific evidence he offers to demonstrate that the earth and everything upon it should be conceived as a living system is far from convincing.

Emphasizing the break from patriarchal, male-centered conceptions of man and nature, many radical feminists have embraced Gaia as an alternative view of ecology that corresponds to many of their movement's goals. Ecofeminist proponents of the Gaia hypothesis view it as a positive movement of social reordering away from theories of domination and toward a more inclusive global consciousness of the relationship between humanity and the overall system of which it is but one facet (Ruether 1992; Zimmerman 1994).

ANTI-ENVIRONMENTAL BACKLASH

Organized campaigns against the modern environmental movement intensified in the past decade. Although many environmental laws receive wide public acceptance, Americans are far from unanimous in their support of recent initiatives. As the number and scope of environmental regulations increase, ecological policies directly affect greater numbers of people, some of whom choose to fight against what they consider to be an abrogation of their rights.

Concerned about the future of the environmental movement, recent articles in the *Utne Reader* (Walljasper 1992) and the *Progressive* (Knox 1992) lament the growth of anti-environmental campaigns. Recent grass-roots campaigns against environmental initiatives are based on the view that when the two issues collide, the economy must be more important than the environment. Indeed, most Americans have always assumed that the environment could be protected without serious personal cost to them. Most Americans still consider themselves environmentalists. Cost does concern them, however, and they feel less urgency about the environment than they did ten years ago (Ladd and Bowman 1995).

The most extensive and organized anti-environmental campaigns focus on land use policy. Although the debate over the best use of public lands dates back to the nineteenth century, the growth of the modern environmental movement escalated tensions over public land policy in the 1980s and 1990s. Farmers, ranchers, foresters, and others who derive income from use of public lands and resources had entirely different ideological and economic concerns from those of environmental activists. These groups joined in the Sagebrush Rebellion, a largely symbolic regional protest movement that advocated the return of public lands to states (Gregg 1991, pp. 155–156, 162–163).

The current "wise-use" movement represents a larger-scale reaction against environmentalist policies. It brings together land rights activists who variously advocate greater protection of property rights and wider latitude in permitting the use of natural resources. The property rights faction argues for greater government compensation for economic losses incurred by governmental regulations. The wise-use contingent organizes against regulations that restrict activities on federal lands. In addition, a number of prominent journalists and scholars who were once considered strong supporters of environmentalism have recently directed unaccustomed criticism against both the mainstream and more radical environmental movements.[31]

Conclusion and Prologue

The election of Bill Clinton in November 1992 signaled the end of twelve years of Republican control of the White House. Since he had been elected with support from major environmental organizations, environmental activists had great hopes that President Clinton would champion significant environmental programs.

In his first two years in office, however, Clinton received mixed reviews for his environmental agenda. Although he did propose policies to facilitate the development of environmentally friendly technologies (Marshall 1993), most environmental activists complained that Clinton had done little to further the goals of the environmental movement. Having devoted most of his energies to developing broad proposals for health care and welfare reform, Clinton sided against most environmental groups on his one other major legislative crusade of 1993, the ratification of the North American Free Trade Agreement (NAFTA). One disappointed advocate worried that Clinton's style "of compromise politics is going to be applied throughout the environmental arena, from old-growth forests to coastal development to waste-treatment plants—in which compromise means nothing less than losing half the battle" (Sale 1993, p. 95).

Anxiety among environmentalists increased with the 1994 Republican congressional victory. Would Republican initiatives reverse established government policies? Worse, did they presage a Congress and President in 1996 both hostile to the environmental movement? What they feared did not come to pass. The Republicans lost their campaign to regain the presidency, and their majority in the House of Representatives was sharply reduced. For the most part their legislative initiatives failed.

In his second term President Clinton has continued to receive mixed reviews. Even Vice President Al Gore, whose public commitment to environmental regulation has been stronger than the president's, failed to gain the plaudits of the more militant environmental groups. Despite their disappointments, however, the environmental movement keeps going from success to success. It has too much generalized support to be stopped at this point and is likely to remain a potent force in politics. After all, despite the vocal hostility of the Reagan administration and the sometimes limited cooperation of other presidents, environmentalists have achieved and consolidated a remarkable number of policy successes in a relatively short time. Even more significantly, the impact of the movement has raised public environmental consciousness. Americans in the 1990s are aware of environmental issues, and significant majorities support, within limits, further regulations to protect the planet. What the future holds for the movement, however, is an issue we will deal with at greater length in the final chapter of this book, when we have more data with which to work. The study of environmental cancer which is described in the next three chapters will provide some of that data.

3
What Is Environmental Cancer?

AMERICANS FEAR CANCER MORE than any other disease. One Gallup poll found that the public overwhelmingly selected cancer as "the worst thing that can happen to you" when presented with a list of illnesses. Heart disease, which actually kills more people each year, was six times less likely to be selected (Blumenthal, 1978).[1] This fear is hardly groundless. Nearly four hundred thousand Americans will die from cancer this year, and it often kills in a slow, painful fashion. Heart disease is widely perceived as a quick, "clean" death, while cancer is seen as debilitating and horrible. The cost of medical care and the impact on the families of cancer patients add to individual suffering and thereby contribute to the widespread fear of cancer.

Real as these worries are, some critics charge that public perceptions of cancer risks are exaggerated. They argue that rational concern has given way to "cancerphobia," the belief that virtually everything we eat, drink, wear, breathe, or touch has the potential to cause cancer (Whelan 1978, 1981; Lichter 1993). Some substances, notably tobacco, clearly deserve their reputation as dangerous carcinogens. But estimates that 90 percent of all cancers are caused by "contaminants placed in the environment by man," as a United States senator once suggested, serve to frighten rather than enlighten.[2] Cancer, as this estimate implies, is a "political" disease because of the debate over just how much of it is preventable. How much cancer is caused by "natural" factors such as diet and sunlight and aging, and how much is "man-made" and therefore avoidable? Even the most advanced scientific research on this subject is still quite uncertain and, as always, subject to revision. But even so, it is possible to summarize recent and ongoing research into environmental cancer.[3]

First, we shall consider the biological and medical understanding of cancer, including the definition of environmental cancer used by cancer researchers. Second, we shall address the relative importance of the various environmental causes of cancer and examine some of the scientific and political controversies over their incidence. Third, we shall review some methodological problems involved in the epidemiology of cancer, along with the uses and abuses of data in the analysis of environmental cancer. Finally, we shall describe the "issue network" of interest groups, bureaucracies, experts, journalists, lawyers, and others involved in cancer prevention and regulation.

We shall not deal with the history of the war on cancer or federal activities in cancer research at earlier periods. That was covered ably by Stephen Strickland (1972) and James Patterson (1987), though we do not necessarily agree with all their emphases. Rather the increased emphasis on environmental triggering of cancer and attempts to prevent such triggering during the 70s and 80s stems at least partly from a decline in the expectation that we can find a cure. Despite the large expenditure of money on seeking a cure, very little progress appears to have been made during that period.

Medical and Biological Models of Cancer: Background and Definitions

Before attempting to define environmental cancer, it is necessary to define cancer itself. This is no simple matter, because there are many cancers. That is, the term is used to refer to some two hundred diseases. A few are easily curable by modern medicine, some are almost chronic illnesses, and others are quickly fatal to virtually everyone unlucky enough to contract them. Moreover, cancers arise at different sites in the body, attack different types of individuals, and are enormously varied when examined closely.

The common characteristic shared by all cancers is that cells begin to reproduce beyond the control of the body. But mere rapid reproduction of cells does not necessarily indicate a cancerous condition. Red blood cells are replaced in healthy bodies in only four months, and white blood cells even faster. The crucial characteristic is that cancerous cells seem to become autonomous of control mechanisms that are normally in force (see Richards 1972; Glasser 1976, pp. 4–5). These uncertainties of definition represent more than an academic exercise, because the way a researcher defines cancer will largely determine the reported incidence of the disease and the fatalities ascribed to it. As our clinical capacity for recognizing various cancers has increased, so have the number of cases and deaths recorded in official accountings. In 1940, for example, a number of diseases previously classified otherwise were determined to be a variety of cancer, thus increasing reported cancer rates (Dr. Hardin Jones, quoted in Moss 1980, p. 34). Definitions, then, can determine data. Many deaths that would have been attributed to other causes in the 1930s are now ascribed to cancer.

Changes in the definition of cancer have followed discoveries in medical research. Cancer was known to physicians even in ancient Greece and Rome, but modern investigations into its causes can be traced to Percival Pott, a British doctor in the eighteenth century who hypothesized—correctly, as was later shown—that the high incidence of cancer of the scrotum among chimney sweeps was due to their regular contact with coal tar.[4] Coal tar's carcinogenic properties were confirmed by two Japanese researchers in 1915 (Sugimura 1986). They showed that it was possible to induce tumors by repeated application of coal tar to the skins of laboratory animals.

In the early 1920s Peyton Rous almost accidentally discovered that tumors could be made to develop more quickly than the Japanese research had suggested. Rous had been painting the ears of rabbits with coal tar when he needed to move the animals. To make identification of specific rabbits simpler, he punched small holes in their ears. Tumors quickly began to appear near the punch holes. This was the first hint that cancer causation might be divided into two phases: induction and promotion.

In the 1940s Isaac Berenblum, one of the great early cancer researchers, elaborated on Rous's work. Berenblum used benzopyrene—a distillate of coal tar and an even more powerful carcinogen—to show that there were separate cancer inducers and cancer promoters. First he painted the skin of laboratory mice with benzopyrene in a dosage that should have been too weak to induce cancer. Then he rubbed the skin with croton oil, which was known to irritate skin but was not itself a carcinogen. Yet the combination of the low dose of benzopyrene and the croton oil produced malignant tumors among the mice very quickly.[5]

In the 1950s Berenblum discovered that exposure to a cancer inducer (also known as an "initiator"), followed by exposure to a cancer promoter would result in cancer even years later. Of course, some substances are so carcinogenic as to cause cancer by themselves, but the initiator-promoter linkage helps to explain why cancer frequently takes decades to develop. Assuming that there are frequently two distinct steps in the process of carcinogenesis, a given cell or group of cells may be transformed into a "pre-cancerous" condition by the initiator and then wait in this altered state for years before an "insult," such as the puncture of the rabbits' ears, or an encounter with a promoter, such as croton oil, makes them actively cancerous.

There is an additional route by which cancer can be caused, in which two or more substances interact synergistically. The result may be to produce more cancers than can be accounted for by adding the carcinogenic effects of each substance individually. For example, it is now well known that exposure to asbestos, when combined with smoking cigarettes, causes many more cases of cancer than can be explained either by smoking or by exposure to asbestos alone. But the impact of synergism on carcinogenesis was discovered after the initiator-promoter linkage was found.

At the time Berenblum and his colleagues revealed the importance of this linkage, it appeared that initiators were few in number and synthetic (man-made) in character. It seemed plausible that this limited list of chemicals could be identified and kept out of human food supplies. This approach was soon reflected in public policy by the Delaney amendment to the food and drug laws. The Delaney clause absolutely banned any food additives that had been found to "induce cancer in man or in animals."[6] Leaving aside the substantially improved capacity of contemporary science to detect minute amounts of chemicals that were undetectable in 1958, subsequent research revealed that many potent carcinogens are naturally present in food, such as aflatoxins and safrole (see Ames 1983, 1992a; Ames, Magaw, and Gold 1987; Ames, Gold, and Willett 1995; and Ames and Gold 1997). But this is getting ahead of the story. The next major advance in medical understanding of cancer grew out of work done on the role of deoxyribonucleic acid, or DNA, in human cells.

The genetic material DNA is identical in every cell in the body, but the expression of genes—which ones are "turned on" and which are "turned off"—differs in brain cells, lung cells, stomach cells, etc. All normal cells share the characteristic of regulated growth. Cells know their place: they don't grow to the point that they interfere with neighboring cells or disrupt the functions of tissues and organs. Cancer cells, on the other hand, are often described as being out of control. One of the steps along the way to that loss of control is alteration of their DNA. Such an alteration, called a "mutation," can be caused by any of a huge number of natural and synthetic chemicals and various forms of radiation. Most mutations have no adverse consequences. It has been estimated that up to ten thousand mutations may occur in the average human cell daily. Rarely, however, a mutation can initiate the process that results in cancer, and for this discussion the terms *mutagen* and *initiator* can be used interchangeably.

The well-known Ames test to determine whether a substance is mutagenic depends on the relationship between mutagenesis and initiation. Since the usual animal tests take two to three years to complete and can cost hundreds of thousands of dollars (see Proctor 1995, pp. 166–171; and Glasser 1976, pp. 11–17), the Ames test has been employed as a rapid and inexpensive screening test to separate chemicals into mutagens and nonmutagens. With regards to carcinogenicity, greater suspicion is attached to mutagens because they generally have the capacity to act as initiators. (Some don't, perhaps because the concentrations of the chemical that would cause initiation and cancer would kill the test animals outright.) Moreover, some mutagens are carcinogenic, indicating that substances that cannot initiate cancer can promote its appearance in animals in which natural biochemical processes or unknown initiators present in animal food or elsewhere have initiated some cells.

Whatever tests are used to judge the carcinogenic potential of a given substance, there is a substantial irony in the position of Bruce Ames. Once identified with those who suggest that a large fraction of cancer is caused by industrial and consumer products, Ames shifted his position in light of later research that emphasized the dominance of natural carcinogens and the minimal, often nonexistent risk of cancer from trace amounts of chemicals of whatever origin.[7]

Sources and Patterns of Environmental Cancer

The term *environmental cancer* is ambiguous, conveying quite different meanings to cancer researchers and lay audiences. This confusion about definitions provides a partial explanation for the sometimes exaggerated charges made against consumer goods, industrial products, pesticides and herbicides, food additives and preservatives, and air and water pollutants, among other things. When most research scientists refer to environmental cancer, they have in mind a wide array of causes, including natural as well as man-made agents. To use the definition provided by the congressional Office of Technology Assessment, environmental cancer refers to cancer caused by "anything that interacts with humans, including substances eaten, drunk, and smoked, natural and medical radiation, workplace exposures, drugs, aspects of sexual behavior, and substances present in the air, water and soil" (Office of Technology Assessment 1981, p. 3).

As Richard Doll and Richard Peto point out, the phrase " 'environmental factors' . . . has been misinterpreted by many people to mean only 'man-made chemicals' " (Doll and Peto 1981, p. 1197). A number of man-made chemicals are indeed carcinogens. As the broader definition of environmental cancer used by these scientists suggests, however, to significantly reduce the incidence of environmental cancers would require substantial changes in life-style. Notably, in North America and Western Europe, many dietary preferences and personal habits such as smoking would need to change to reduce cancer rates.

In the population of the United States, the sites in the body in which cancers are most commonly found include the lungs, the colon, the skin, the breasts (in women), and the prostate (in men). Among sites in which cancer is less common in Americans but not rare are the bladder and the pancreas; and there are a few sites in which the disease is quite rare but nonetheless occurs, such as the liver. The significance of the uneven distribution of cancer in the body is that the same patterns are not found uniformly around the world. In Japan, for example, breast cancer is far less common than it is in the United States, but stomach cancer rates are considerably higher. In certain poor countries, liver cancer is much higher than in the United States, while other forms of cancer are less common. Cancer of the esophagus is fairly common in China and the Caribbean but is uncommon both here and in Western Europe.[8]

The disease rates of people who migrate from one society to another provide what might be called natural experimental data. For example, Japanese women who have moved to the United States show increased rates of breast cancer, but not as high as their daughters and granddaughters, who are more fully integrated in American culture. Moreover, as Western diets become more common in Japan, Japanese girls are reaching menarche at younger ages, increasing the length of exposure to natural estrogens; as a result, breast cancer rates in Japan are increasing. Similarly, it is possible to compare African-Americans who have lived in North America for generations with their ethnic cousins in Africa. The results of such comparisons show striking differences in the types of cancer to which each group is most susceptible. Thus, blacks in the United States suffer much more often from lung cancer, but African blacks suffer more frequently from liver cancer.[9]

There are even notable differences between subcultures within the same societies. Mormons in Utah and other Western states, for example, are much less susceptible to lung cancer than their non-Mormon neighbors (Doll and Peto 1981, p. 1200). This finding is highly suggestive, since tobacco use is religiously proscribed among Mormons. Of course, tobacco has been indicted by study after study as the greatest single cause of cancer in the United States, particularly lung cancer.

The utility of national and migration data is suggested by the findings on lung cancer among the Mormons. Why, then, do Japanese men and women suffer from relatively high rates of stomach cancer, while Japanese women have relatively low rates of breast cancer? Many factors are probably involved, but researchers have pointed out that in Japan highly salted fish is a major part of the diet. A favored method of preparation involves broiling, which results in consumption of substantial amounts of complicated carbon compounds found in browned and blackened foods. Such compounds are known to be mutagenic and are thought to have carcinogenic effects. On the positive side, a diet of fish is lower in fat than the typical American diet, and some studies have pointed toward excessive fat as a suspected agent of breast cancer, though more recent studies have questioned the connection.

Salted meats and fish are uncommon in the American diet, and Americans are less likely than are the Japanese to eat broiled (blackened) food. Though the consumption of such foods is increasing in the United States somewhat, in countries like Japan and China these foods are far more heavily salted or cured then they are here. Probably for this reason Americans boast relatively low stomach cancer rates (as compared to the Japanese), but higher cancer rates in other organs and tissues including the breast (see Richards 1972, pp. 122–123; Office of Technology Assessment 1981, pp. 76–84; Whelan 1994, pp. 265–266; Fuchs and Mayer 1995; de Meester and Gerber 1995).

Estimates of the Environmental Causes of Cancer

As these examples suggest, when estimates are made of the proportion of cancer caused by the "environment," tobacco and diet lead the list of suspected carcinogens. Indeed the best estimates suggest that diet (taken as a whole) probably contributes slightly more to the incidence of cancer than does tobacco, though they each probably account for almost one third of cancers. To be sure, industrial and occupational exposures to carcinogens are are also significant but fall rather far down the list, representing a relatively small fraction of all environmental cancers. The estimates of experts vary concerning the proportion of cancers caused by various suspected carcinogens, but a number of scientists have come to similar conclusions.[10]

The most exhaustive epidemiological study on cancer was done by Doll and Peto in connection with their report for the congressional Office of Technology Assessment in 1981.[11] They estimated that tobacco was responsible for about 30 percent of all American cancers, allowing that the "range of acceptable estimates" was 25–40 percent. They estimated that diet was responsible for another 35 percent, infection perhaps 10 percent, reproductive and sexual behavior about 7 percent, occupational hazards about 5 percent (the range of acceptable estimates was 2–8 percent), geophysical factors (such as sunlight) 3 percent, alcohol 3 percent, pollution 2 percent, medicine and medical practices 1 percent, and food additives and industrial products less than 1 percent each. The most recent data from national surveys for 1987–1991 show some variation in specific cancer incidence, but the patterns remain largely as they were in the early 1980s when the baseline epidemiological studies were conducted (Doll and Peto 1981, pp. 1220–1256).[12]

Doll and Peto's estimates were anticipated by two independent estimates by distinguished cancer researchers. In 1977 Ernst Wynder (the first scientist to clearly demonstrate that cigarettes were carcinogenic) and Gio Gori used somewhat different categories to estimate the causes of cancer in the United States. They produced separate estimates for men and women. Wynder and Gori estimated diet to be responsible for 57 percent of cancers among women and 40 percent among men, tobacco to be responsible for 25 percent among men and 8 percent among women (and the combination of tobacco and alcohol another 4 percent among men and 1 percent among women), sunlight and radiation 8 percent, occupational hazards 4 percent among men and 2 percent among women, and hormones 4 percent among women only, leaving 16 percent among men and 20 percent among women either congenital (inherited) or of unknown origin (see Wynder and Gori 1977).

In 1979 John Higginson and C. S. Muir estimated that in the Birmingham region in England, tobacco might account for 30 percent of cancers among men, "life-style" factors such as diet and sexual behavior another 30 percent, sunlight 10 percent, tobacco and alcohol in combination another 5 percent, occupational hazards 6 percent, miscellaneous causes (including inheritance) 4 percent, and 15 percent of unknown origin. Among women, Higginson and Muir estimated that fully 63 percent of cancers could be attributed to "life-style" factors, with the other proportions roughly the same as for men except that tobacco was much less significant—only 7 percent—and occupational hazards were somewhat lower, 2 percent instead of 6 percent.[13] These three sets of estimates by scientists who are among the most respected in the field are summarized in table 3.1.

Table 3.1
Estimates of Environmental Causes of Cancer

	Doll and Peto	Wynder and Gori		Higginson and Muir	
	1981	1977		1979	
	United States	United States		Birmingham, England	
	Men and women	Men	Women	Men	Women
Diet	35	part of "life-style"		part of "life-style"	
Tobacco	30	28	8	30	7
Alcohol	3	4[a]	1[a]	5[a]	3[a]
Life-style	—	—	—	30	63
Sunlight/Radiation	3	8	8	11	11
Reproductive/Sexual Behavior	7	—	—	part of "life-style"	
Occupation	4	4	2	6	2
Pollution	2	—	—	—	—
Medicine/medical practices	1	—	—	1	1
Infection	10?	—	—	—	—
Miscellaneous[b]	1	—	4	2	2
Congenital/ Unknown	?	16	20	17	13

[a] *Wynder and Gori (1977) and Higginson and Muir (1979) list this factor as the combined effects of alcohol and tobacco.*

[b] *Doll and Peto (1981) list both food additives and industrial products at less than 1%; Wynder and Gori include the effects of hormones on women only; Higginson and Muir list radiation and "iatrogenic" factors.*

Each of the major environmental factors associated with cancer deserves separate consideration. These are diet, tobacco, alcohol, occupational hazards, sunlight and radiation, sex, drugs, food additives, pollution and pesticides, and unknown causes.

DIET

The link between cancer and dietary habits deserves closer attention than it has so far received. The pattern of high rates of breast and colon cancer in the United States and Western Europe, as compared to Third World countries, has generally been explained as a function of the amounts of fat and fiber in the diet. In most developed nations, except Japan, diets tend to be quite high in fat. It may be that the body produces additional hormones in response to these high levels. Those hormones may either interact with other bodily substances to produce carcinogenic effects, through growth promotion, or they may suppress the body's immune system. If either of these hypotheses is right, fat would have to be regarded as a carcinogen.

The American diet is high in fat but low in fiber. It has been hypothesized that potentially carcinogenic body waste products may remain in the colon and rectum overly long. Diets high in fiber, such as those typically found in poor countries, cause stools to be passed through the colon more quickly, and it has been suggested that high-fiber diets might reduce the cancer rate. Indeed the best evidence appears to indicate that diets high in fat and low in fiber predispose individuals to colon cancer. A high fat diet is also implicated in breast cancer and prostate cancer. The dietary fat–cancer linkage is supported by many clinical trials on both animals and humans. A diet in which no more than 10 percent of the calories are from fat and which includes at least forty grams of fiber daily will prevent up to 35 percent of all cancers.[14]

Less commonly discussed but quite important in understanding the role of diet in cancer are carcinogens that occur naturally in food. Perhaps the best known of these is aflatoxin, a powerful carcinogen that is produced by a mold that can grow on certain nuts, notably peanuts. There are many other carcinogens and suspected carcinogens among foods, as Bruce Ames has shown. Fortunately, Ames and Gold suggest, there may also be natural anticarcinogens in certain foods (Ames 1983).[15] Certain cooking procedures may also pose hazards for cancer. As noted, the Japanese preference for broiling their fish (and other foods) is one strong suspect in this category.

TOBACCO

The use of tobacco, particularly by smoking, is overwhelmingly the largest single avoidable cause of cancer in the United States and probably the world. Although its dangers are widely recognized today, they were not always obvious. The habit of cigarette smoking spread tobacco use throughout the United States during the first half of the twentieth century. By 1935, two-thirds of adult American males and about a quarter of adult females under the age of forty were smoking cigarettes. Men tended to take up smoking earlier than women and were more likely to smoke at some point in their lives. This pattern is now reflected in substantially higher rates of lung cancer among men than women, though the long period between initial tobacco exposure and the development of cancer means that lung cancer rates among women will almost certainly continue to rise in the near future.

Evidence of health hazards from smoking began to appear as the habit spread. The research that first clearly pointed to smoking as the overwhelming cause of lung cancer was done by Ernst Wynder while he was a graduate student working with Evarts Graham at the Washington University Medical School in Saint Louis. Graham and Wynder studied almost seven hundred lung cancer patients at the medical school hospital. Ninety-four percent smoked cigarettes, 3 percent smoked pipes and 3 percent cigars. Soon after these results were published in 1950, two British researchers, Richard Doll and Bradford Hill, found similar patterns among lung cancer patients in twenty London hospitals.

These early studies were not enough to convince the public or even the skeptical medical profession that smoking was a leading cause of lung cancer. But further research, including a huge epidemiological study by the American Cancer Society, in conjunction with continuing research in England by Doll and Hill, clearly demonstrated the connection. In June 1957 the *New York Times* ran a front-page story titled "Cigarette Smoking Linked to Cancer in High Degree" (Schmeck 1957). By 1964 the Surgeon General had issued the famous report warning of the health dangers associated with smoking (summarized from Doll and Peto 1981, pp. 1220–1224; Fischer 1982, pp. 8–9; Whelan 1978, chap. 3). Many parallel reports have appeared since, extending and amplifying the same basic message that smoking is extremely hazardous to human health.

This message appears to have reached the public and created substantial awareness of smoking hazards. In fact, some individuals seem to overestimate the risks, if one considers the scientific evidence (Viscusi 1992). Yet despite public awareness of the risks, a

Table 3.2
Lung and Respiratory Cancers Versus All Other Sites[a]
for Selected Years

Year	Lung/Respiratory	All Other Sites
1935	5,812	58,709
1945	11,818	60,800
1955	19,495	57,938
1965	26,766	57,390
1970	30,683	55,282
1975	32,748	53,356
1978	33,816	52,538

[a] *Incidence rates for males, per one million population. For women, there is a steeper decline in cancers outside the respiratory system, but lung and respiratory cancers increase rapidly beginning in the mid-sixties.*

significant, though declining, minority of the populace continues to smoke tobacco.

The greatest change in public attitudes toward smoking concerns the willingness to take actions to curb smoking behavior. By the 1980s enthusiasm for restricting smoking in public places had become widespread. This support has led in the 1990s to smoking restrictions in almost every locale. Imposing smoking restrictions has become more acceptable politically as the fraction of smokers in the population has diminished to about one-third (Viscusi 1992, p. 6).[16]

The most heated contemporary disputes concerning tobacco revolve around three issues: the health hazard of secondhand smoke, the addictive properties of nicotine, and the fiscal responsibility of tobacco companies for tobacco-related health costs.

Secondhand smoke, or environmental tobacco smoke, is smoke exhaled by cigarette, cigar, or pipe smokers that is subsequently inhaled by nearby nonsmokers. Recent studies have found that secondhand smoke is a potential carcinogen to people who are extensively exposed to it over a long period of time, though this is still a matter of some dispute (see Procter 1995, pp. 105–110; Devesa et al. 1995).

In recent years the FDA has been moving forward in an effort to further regulate tobacco as an addictive drug. According to the FDA's proposed rule, "Nicotine in cigarettes and smokeless tobacco has the same pharmacological effects as other drugs that the FDA has traditionally regulated" (61 Federal Register 168, 44661, 1966). The goal of the FDA is to attempt to regulate tobacco as a nicotine "delivery device." The nicotine itself is not considered a carcinogen. The FDA intends to establish national regulation that would measure the nicotine content in tobacco, restrict tobacco sales to minors, and further limit tobacco advertising, which is already banned from television and radio.

During the early 1990s the number of class-action law suits against tobacco companies escalated. Initially, individuals tried to obtain compensation for illnesses and deaths caused by smoking. Court decisions, however, have been mixed. Some rulings by judges have implied that it might be difficult for smokers, ex-smokers, or third parties to be compensated under rules governing liability (*Economist* 1996, May 11), while others have been more encouraging.[17]

In 1997 a host of state attorneys general filed suit against tobacco companies to recover money spent by governments in treating tobacco-related diseases. Forty states filed claims, and the tobacco companies have entered into settlement agreements with Texas, Florida, and Mississippi. Because of the threat of further litigation the tobacco companies have sought a deal with all the state governments whereby the companies would pay specified sums (in the billions of dollars each year) and reduce attempts to increase tobacco sales in prescribed ways. In return, the companies would receive immunity against further state or class-action suits. Because of its scope the agreement would require an act of Congress.

Although the agreement is historic, it faces opposition from both smoking and anti-smoking advocates. For example, David Kessler, the FDA's former director, and former surgeon general C. Everett Koop have opposed the agreement in part because they believe it heavily restricts possible regulation by the FDA to reduce the number of adolescents who take up smoking as well as to help current smokers quit. However, several southern Republicans, whose states benefit from tobacco revenues, dominate many of the relevant subcommittees. Not surprisingly, Congress as a whole has been unable to come to a consensus on the tobacco agreement.

It appears, though, that a turning point has been reached: smoking and the production and sale of cigarettes in the United States will be increasingly subject to regulatory constraints. Whether or not this changes the behavior of current and potential smokers remains to be seen.

The position of the tobacco companies has become weaker because of heightened safety consciousness by an increasing number of nonsmokers and ex-smokers who find tobacco smoke unpleasant and are convinced that it poses a danger to their health. It also has been blunted by the publication of books, articles, and newspaper stories which appear to demonstrate that tobacco companies had manipulated the amount of nicotine in cigarettes and cigars so as to increase their addictiveness and had lied to the public as well as to Congress.[18] Whether or not press or book coverage was completely fair, tobacco companies found themselves facing an increasingly hostile public and political figures who were no longer willing to accommodate their requests despite substantial funds spent on lobbying.

ALCOHOL

Some research suggests that alcohol may not by itself be carcinogenic. One of the problems with research in this area, however, is that heavy drinkers are frequently also heavy smokers. Alcohol and tobacco appear to have a synergistic carcinogenic effect. When drinkers also smoke, their chances of developing cancer—mainly cancer of the mouth, larynx, esophagus, and liver—significantly increase. For example, those who smoke two packs of cigarettes a day increase their chances of oral cancer by 143 percent. If they also imbibe two alcoholic drinks a day, however, their chance of developing oral cancer increases by more than 1,400 percent above that of nonsmokers.[19] It should be added that the National Research Council's 1996 study of dietary carcinogens concluded that alcohol alone may be a cause of some cancers, including liver cancer (p. 281).

OCCUPATIONAL HAZARDS

There are real dangers from carcinogens in the workplace. All the major epidemiological studies agree on this, though the estimates of the dangers posed (as compared to those associated with smoking or diet) are fairly low. Higginson and Muir estimated that as many as 6 percent of all cancers in men might be caused by occupational exposures to carcinogens. These researchers, along with Wynder and Gori, estimated that 2 percent of all cancers in women could be traced to occupational causes. Doll and Peto, providing a unisex estimate, split the difference, attributing 4 percent of all cancers to workplace exposures.

One out of twenty-five cancers each year, while far smaller than the rate of one out of three cancers associated with smoking, represents a great deal of human suffering—twenty thousand deaths owing to occupational cancer. So the strong position taken by labor unions on the issue of occupational cancer keeps faith with their members. Moreover, in a number of instances, management has unnecessarily exposed workers to known carcinogens or, at the very least, displayed studied ignorance when evidence was available of the dangers of cancer in the workplace (Whelan 1978, chap. 10).

One example of corporate irresponsibility involved certain chemicals used in the production of artificial dyes. These were shown by Wilhelm Hueper to be carcinogens in the mid-1930s, but they continued to be used for years in the chemical industry. Hueper was a pioneer researcher on occupational cancer who went on to head the Environmental Cancer Section of the National Cancer Institute. As a result of his early experiences, Hueper became convinced that industry could not be trusted (see Hueper 1942; and a review of Hueper's work by Proctor 1995, pp. 38–48). We will return to this theme; for now, the point is that at least some of the current difficulty for industry can be traced to earlier unconscionable behavior.

At the same time, no constructive purpose is served in exaggerating the real dangers of carcinogens in the workplace, as Joseph Califano did in 1978. Califano, then secretary of Health, Education and Welfare, told a labor audience that between 20 percent and 38 percent of all future cancers would be caused by occupational exposure to carcinogens. Califano's speech came during a time when press coverage of environmental cancer was intense. The attention—particularly in a 1976 CBS documentary—probably contributed to focusing official attention on environmental cancer (Proctor 1995, pp. 64–69).

In 1977 officials at OSHA were joined by three other regulatory agencies—the Consumer Product Safety Commission, the FDA, and the EPA—to review policies on risk, particularly cancer risk. They issued a report through the National Cancer Institute, the National Institute of Environmental Health Sciences, and the National Institute for Occupational Safety and Health entitled "Scientific Bases for Identification of Potential Carcinogens and Estimations of Risk." The paper, in a slightly different version, was titled "Estimates of the Fraction of Cancer in the United States Related to Occupational Factors." For simplicity, we will refer to it as the OSHA paper, reflecting its parent agency.

The OSHA paper originally appeared in the *Federal Register* and was later published in the *Journal of the National Cancer Institute,* but it did not undergo peer review prior to publication. The authors were prominent government scientists serving in administrative posts in their agencies (Occupational Safety and Health Administration 1978; for more, see Doll and Peto 1981; Rushefsky 1986). The estimates in the OSHA paper were and are much larger than any others calculated by serious scholars or scientists. Its estimates of occupational cancer risk, which were publicized in Califano's speech, set off alarms in the labor movement both in the United States and in other developed countries. The OSHA paper was widely received as reflecting scientific work, but the administrators who signed it did so following a convention which meant that none of them took personal responsibility for the document. The authors were listed alphabetically, with no specific attribution of who was responsible for the research.[20] The reasons for this procedure became apparent after their estimates were strongly criticized by numerous experts in environmental carcinogenesis.

The methods used in the OSHA paper to calculate future rates of cancer from occupational sources were at once simple and misleading. Their essence was to take the highest reported exposure of workers to a given carcinogen, extrapolate that to all workers who had ever come in contact with the substance, even if such contact was for a short time and with protective regulations and procedures in effect, and assume that *all* workers in a given industry would be subject to that level of risk. After reviewing the calculations for the nickel-refining industry, Doll and Peto described this procedure as one that "might fairly be described as a confidence trick" (Doll and Peto 1981, p. 1305). Similar exaggerated estimates for workers exposed to asbestos during World War II account for about half the expected increase in occupationally caused cancer in the OSHA paper. That is not to deny that former asbestos workers are suffering from cancer caused by their wartime exposure, but the magnitude of the problem was exaggerated significantly in the OSHA paper (Doll and Peto 1981, pp. 1306–1308; Brodkin et al. 1996; Renke and Rosik 1994).

If the scientific questions posed in this controversy seem to have been clearly resolved, the same cannot be said for the political questions. These estimates of the potential for occupational cancer constituted another round in a lengthy struggle between labor and management. These heavyweights in American politics have slugged it out over a wide variety of workplace safety issues (see Kelman 1980, chap. 7). But rising rates of occupational cancer (or projections of them) did not cause union leaders to seek regulations from friendly bureaucrats or to ask for hearings by sympathetic members of Congress. Instead, the OSHA paper estimates were generated from within government, and these estimates then created a demand for action by interest group leaders who perceived their members to be endangered. This pattern of policy initiation, from government to constituencies and then back again as new policies, has been called "policy as its own cause" (see Wildavsky 1979, chap. 3).

The OSHA imbroglio demonstrates how some committed activists or individual policy entrepreneurs, such as Ralph Nader or Richard Viguerie, have repeatedly defeated groups traditionally thought to be among the most powerful in American society. Efforts to promote auto safety, control pollution, and legislate fuel economy, as well as the battle over abortion, are among other recent examples of a shift away from material toward ideal interests in our politics, to use the distinction developed by Max Weber.[21] But the road to success is not always smooth. To obtain the passage of the environmental legislation one champions (and to advance subsequent regulation) often requires support from groups which, at first blush, have no economic stake in them—for Nader, this has included trade unions, college students, and trial lawyers. That is not to say that union leaders were hoodwinked. Long-simmering conflicts with management over workplace safety made labor leaders ready believers once the OSHA paper estimates were published. The politics of occupational cancer, however, often involve hidden agendas.

SUNLIGHT AND RADIATION

All the major epidemiological studies agree that sunlight and radiation may cause cancer, though estimates of their impact vary somewhat more than estimates regarding the incidence of occupational cancers. Higginson and Muir attribute 11 percent of all cancers to the combined effects of sunlight and man-made radiation (such as X-rays). Wynder and Gori put the figure at 8 percent. Doll and Peto, who assign medical radiation to a separate category, ascribe a lower proportion (3 percent) to "geophysical factors."

Of all the significant causes of cancer, sunlight is probably the most difficult to eliminate, at least among light-skinned persons, who are most at risk from solar radiation. Skin cancer has notably increased in the United States since suntanning became popular earlier in this century. Fortunately, sun-caused cancers are usually treatable if diagnosed early enough.[22]

More worrisome, although less widespread, are cancers caused by man-made radiation, mainly X-rays and other radiation used in medical diagnosis and treatment. Doll and Peto estimated such exposures to cause about one-half of one percent of all cancer deaths in the United States, or about two thousand deaths a year (Doll and Peto 1981, p. 1252). At one time, the dangers associated with medical uses of radiation were not appreciated. Medical workers who administered X-rays in the early days of the technology suffered from disproportionately high rates of cancer. As the dangers became clear, however, controls were imposed on the amount of radiation to which any medical worker could be exposed (Whelan 1978, pp. 115–119; Whelan 1994, pp. 189–90; Miller 1995). There are obvious and difficult choices to be made when patients need many X-rays or other forms of medical radiation. (Paradoxically, many of these patients already have cancer, for which they are receiving radiation treatment.) In such cases, the cure may pose potentially serious risks later. The dangers of medical radiation are well respected, however, so the risk of initiating a new cancer by radiation treatment is carefully weighted against the benefits the therapy provides by curbing the present disease.

SEXUAL DEVELOPMENT, REPRODUCTIVE PATTERNS, AND SEXUAL PRACTICES

Doll and Peto estimated that aspects of sexual growth, development, and behavior cause roughly 7 percent of all cancers, though this might now be slightly higher, mainly due to recent increases in reported cases of prostate cancer occasioned by improved diagnosis. Higginson and Muir included this factor in their general "life-style" category. Cancers associated with organs which are involved, even though peripherally, with sexual and reproductive functioning have been more common among women than men, mainly because breast cancer is included under this category, but in recent years, as diagnosis of prostate cancer has improved, the rates for it have edged ahead of breast cancer rates (Devesa et al. 1995, pp. 175–179).

Breast cancer appears to be related to diet, particularly high-fat diets, though recent evidence has brought the connection into question and many other causes are certainly important. The usual uncertainty in ascribing specific causes applies here, but the older a woman is when she bears her first child, the greater her risk of breast cancer. Childless women are at increased risk, particularly when compared to those who have their first child when they are fairly young. Nuns, for example, have relatively high rates of breast cancer. Among mothers, those who breast-feed their babies may have a reduced danger of breast cancer, but the difference is very slight. There is also a pattern of family inheritance of risk. Women with close female relatives who have had breast cancer are two to three times more likely than other women to contract the disease.[23]

Cancer of the cervix seems to be a disease in which a virus or other communicable agent may be involved. While nuns have higher breast cancer rates than sexually active women, they have lower rates of cervical cancer, and prostitutes have substantially higher rates of cervical cancer than do other women. Women who have multiple sex partners clearly increase their risk of cervical cancer. These findings strongly suggest that some contribution from males may be involved in the carcinogenesis of cervical cancer. To some degree, then, cervical cancer can be regarded as an indirect form of venereal disease, but fatalities from it have been declining. Periodic screening for early signs of atypical cells, usually by means of the Pap smear test, may be helping to reduce the death rate.

Among men, the most important cancer of the reproductive system is cancer of the prostate, the incidence of which has been increasing rapidly in the past two decades, probably as a result of better diagnosis. It is a prominent cause of death among older men. Rates for cancer of the testes and penis in the United States are quite low, though the rate for testicular cancer is rising in white men. Cancer of the prostate is notably higher among blacks than whites in this country, but it is less common among black Africans than among white Americans. Again, fat in the diet seems suspect, though hardly proven.[24]

At the time when the major epidemiological studies were conducted, acquired immune deficiency syndrome, or AIDS, had not yet become statistically visible, therefore, any cancers associated with AIDS were not included in the calculations made by Doll and Peto, Higginson and Muir, or Wynder and Gori. The inclusion of statistics on Kaposi's sarcoma and other AIDS-related cancers would increase the fraction of cancer associated with sexual behavior, though such cancers might also be included under those associated with infection.

DRUGS

Certain drugs used in medical treatment may induce cancer. Among the major epidemiological studies Doll and Peto, and Higginson and Muir, estimate that 1 percent of all cancers probably originate in medical treatment, mainly meaning drugs, though medical uses of radiation are also included. Illicit drugs such as marijuana or cocaine were not discussed in this context. There may be parallels between smoking tobacco and smoking marijuana, or between using snuff and "snorting" cocaine. But illicit drugs are beyond the range of research on carcinogenesis for now. Probably the best-known cancer-inducing drug to be widely used in the United States was diethylstilbestrol (DES). DES was commonly prescribed to pregnant women in the late 1940s and 1950s as a preventive against miscarriage. A number of daughters born to women who had taken DES developed a rare form of vaginal cancer in their late teens or twenties. DES apparently acted as a carcinogenic initiator when the fetus was at an early and vulnerable stage of development. The hormones involved in the process of development some two decades later apparently acted as promoters, resulting in cancer for daughters of DES mothers. The risk is not so great as once feared, however—probably no more than 1.4 persons developed this cancer per thousand persons exposed to DES. So far, at least, evidence of parallel problems—say, increased cancer of the penis or testes—for sons of DES mothers is scattered and inconclusive (see Doll and Peto 1981, pp. 1252–1253; Whelan 1994, pp. 210–214; Giusti, Iwanmoto, and Hatch 1995; Sharpe and Skakkebaek 1993).

A variety of other medications in the United States were later implicated as carcinogens, including anabolic steroids and arsenic. Other drugs are now listed as suspected carcinogens, such as phenobarbital and progesterone (Office of Technology Assessment 1981, pp. 96, 98–99). Why are such drugs used? The problem is that these drugs are often the only available means to treat serious illnesses, and the risk of developing cancer is probably too small to outweigh the benefit from treating the existing illness.

No discussion of drugs as potential carcinogens would be complete without mention of oral contraceptives. While the matter is still a subject of controversy, some findings from current studies indicate that both white and black women show an increased risk of breast cancer at young ages with moderate or long-term use of oral contraceptives. The use of oral contraceptives also seems to promote the early development of breast cancer in certain populations of women, particularly those in younger age groups (Mishell 1994; Palmer, J. R. et al. 1995). On the other hand, other studies suggest that use of oral contraceptives may slightly reduce the risk of cervical and ovarian cancers. Other beneficial effects have also been identified. As with the dangers, however, any possible benefits regarding cancer risk are so uncertain as to constitute an inadequate basis for deciding whether to take oral contraceptives or not (Whelan 1994, p. 215).

Politically, potential carcinogens in drugs remain a highly charged subject. As James Q. Wilson has noted, "whatever its incidence, whatever its harm compared to other illnesses or injuries, cancer has acquired a position in the public mind—and thus in political discourse—that subordinates almost every other consideration to its prevention" (Wilson 1980, p. 376). Political incentives for the FDA clearly point in the direction of energetic regulation where even the hint of cancer-causing potential is found in a medicine. This approach is certainly responsive to public opinion, but there are times when difficult choices must be made. When the FDA reviews drugs that have potential carcinogenic effects but also may have unique properties for combating fatal or disabling diseases—certain AIDS drugs come to mind—the predicament posed by an absolute standard is presented starkly.

FOOD ADDITIVES

Food additives as a supposed source of cancer have received a great deal of attention, but the evidence against them is actually quite weak.[25] The major epidemiological studies estimate that food additives make only a negligible contribution to cancer rates; certainly they are responsible for less than 1 percent of all cancers (see Doll and Peto 1981, pp. 1235–1237). Indeed, some additives may even help to protect against certain types of cancer. For example, two commonly used food preservatives, butylated hydroxytoluene (BHT) and butylated hydroxyanisole (BHA), are antioxidants, as are vitamins C and E, and it is possible that declining rates of stomach cancer in the United States are partly due to such additives and vitamins (National Research Council 1996). The evidence on this issue, however, is not strong, though it is clear that the use of additives has reduced the incidence of botulism, which can be life-threatening.

Some food additives used in the past were shown to be carcinogenic. In the United States these include a food dye used to color butter yellow during periods when its natural color was lighter, as well as the additive thiourea. Red dye number two, another food coloring, was banned in 1976 after two Russian studies suggested it might be a carcinogen. Numerous studies in the United States had suggested otherwise, but pressure on the FDA at the time was intense (see Whelan 1978, p. 185). In the case of purely cosmetic additives, such as food dyes, a very cautious approach is only prudent. However, banning one food coloring may lead to the introduction of another posing even greater risks.[26]

When food additives serve to protect public health, a different and more difficult calculus becomes necessary. This latter category includes the controversy over nitrates and nitrites, which are added to certain meats to protect against botulism, a deadly food poisoning. Nitrites used in food do contribute to the total quantity of carcinogenic nitrosamines consumed by Americans, though to a limited degree. Doll and Peto report that some 90 percent of such compounds that reach the stomach come from vegetables (because they naturally occur in such plants as spinach, celery, and lettuce) and human saliva. They also conclude that nitrites in food "must be held responsible for some cases of cancer," though the number is very small (Doll and Peto 1981, p. 1237; National Research Council 1996, pp. 149–50). Here we confront a genuine dilemma. Some small number of cancers are caused by nitrites in cured meat, but people want to eat bacon and ham, not to mention hot dogs. Without nitrites, many more people eating these meats would sicken or die from botulism poisoning than would contract cancer from the nitrites.

Unlike most proposed bans of suspected carcinogens, there was a major public debate over proposed limits on saccharin as a carcinogen, in part because of its use as a sweetener for diabetics who cannot have sugar in their diets. At one point, the FDA even proposed making saccharin a prescription drug. But this does not address the main issue for most people who use saccharin or other artificial sweeteners. They are trying to reduce calories while maintaining sweetness in certain foods. There is weak evidence that saccharin is carcinogenic. It may slightly increase the risk of bladder cancer, but the popular reaction to the announcement of its banning in 1977 was to buy up existing supplies. Cases such as this suggest the outer limits of regulation of food additives, at least when popular items are threatened.[27] As the failure of prohibition during the 1920s suggests, there are limits to the social controls that government can impose on citizens in free societies. In democracies, even health authorities work within limits set by popular preferences.[28]

POLLUTION AND PESTICIDES

Of the major cancer studies, only Doll and Peto specifically estimate a proportion of all cancers (2 percent) caused by pollution of various kinds, including pesticides. Pollution appears to be included in Higginson and Muir's "life-style" category. Air pollution, including an increase in airborne carcinogens, certainly worsened during the first fifty or sixty years of the twentieth century in most American cities, while lung cancer has dramatically increased. Is there a cause-and-effect relationship here?

If no one in the United States smoked and cancer rates had nonetheless increased, then a case might be made for the impact of air pollution on cancer deaths. All serious studies of lung cancer causation, however, have concluded that smoking is the overwhelming cause of the disease. Lung cancer rates are higher in cities than in the countryside because more people smoke in the cities, though it is highly probable that air pollution has exacerbated respiratory damage among urban smokers.

Significant air pollution, such as that found in Los Angeles in the summer, may be damaging to the health of those exposed to it, especially those suffering from respiratory diseases such as asthma. Even this, however, is uncertain (Fumento 1997). In any event, calculations have been made to compare breathing the most polluted urban air with smoking, but they show that damage to the lungs from such pollution is nowhere near that from the average smoking habit. Indeed, secondhand tobacco smoke in a closed room has more effect on the lungs of nonsmokers than does air pollution under all but the most extreme conditions. Nevertheless, it is true that air pollution is independently related to mortality rates, especially fine particulate pollution, including sulfates.[29]

Pollution of water by carcinogens has the potential to damage public health in three ways. The direct consumption of water contaminated by possible carcinogens could be a potential source of increased cancer risk. The consumption of fish or shellfish taken from the contaminated water could increase cancer risk because some pollutants concentrate in their flesh. And, less likely, water containing carcinogens from pesticides or other chemicals may be used to irrigate crops.

Recent research has suggested that there are many carcinogenic contaminants in drinking water. However, this finding is largely a function of the increased sensitivity of measuring devices, which can now identify substances in water down to parts per trillion (Office of Technology Assessment 1981, pp. 92–93; National Research Council 1996, pp. 132–138). Significant contamination of drinking water in the United States seems to derive mainly from asbestos. Some of the asbestos in water supplies comes from old pipes made of concrete and asbestos. Some comes from ground water passing through natural rock formations that include asbestos (Doll and Peto 1981, pp. 1248–1249). Chlorination of water could hypothetically increase cancer risks, but this has not yet been shown, and there are very significant public health benefits from chlorination, which kills many types of pathogenic bacteria. Water fluoridation seems to have no effect or a modest ameliorative effect on cancer rates.[30]

There have been highly publicized cases in which people became ill or even died after eating fish or shellfish taken from polluted water. Perhaps the most notorious of these occurred in Japan. The pollutant in that case was mercury, which is a poison but not precisely a carcinogen. Certain contaminants in food will lodge in human tissue and remain there for long periods of time, so this "food-chain" linkage might yet result in increased tumors. DDT, for example, is anticipated to cause liver damage when concentrations become high enough, though increased liver cancers have not yet been seen in areas where DDT was used before it was banned (Havender 1983a; Doll and Peto 1981, p. 1250; Wildavsky 1995).

Potential carcinogens in agricultural water supplies have also received some attention. It has been feared that through the food chain, humans would build up unacceptable levels of carcinogens or other poisons. This does not appear to be happening so far, but the use of pesticides and herbicides in agricultural areas has certainly increased in recent years, so the possibility bears monitoring (Office of Technology Assessment 1981, pp. 92–93, 108; National Research Council 1996, pp. 132–138).

The potential carcinogenic effects of pesticides applied to crops that will be eaten by humans is one of the most sensitive subjects for government regulatory agencies. As a result of animal tests, the pesticides DDT, aldrin, dieldrin, and ethylene dibromide (EDB) have been banned in the United States. There are real risks from pesticides, particularly for individuals who are regularly exposed to strong concentrations, such as farmers and farm workers, but the case against DDT as a carcinogen in the general population appears to be weak (Havender 1983a, pp. 10–11; Wildavsky 1995). However, DDT received a great deal of attention due to its damaging impact on some wildlife populations.

In 1984 public reports focused on EDB, a compound used as a pesticide on grain and as a soil fumigant against some destructive worms. Studies at the time showed that in high-dose concentrations EDB does indeed induce cancer in laboratory animals such as mice and rats. The concentrations of EDB to which humans were exposed, however, were extremely low; at those levels, it is not at all clear that EDB has carcinogenic effects (Whelan 1994, pp. 201–203). Nonetheless, EDB was banned from use on food crops. Among the difficulties this caused was increased susceptibility of many grains to the growth of aflatoxin. As Bruce Ames and his colleagues have noted, aflatoxin is a natural carcinogen (that is, it is produced by a mold, absent intervention by humans) and unlike EDB is a human carcinogen (Ames, Magaw, and Gold 1987). Moreover, prohibited from using EDB, farmers quickly switched to other pest-control chemicals, particularly methyl bromide, which appear at least as carcinogenic as EDB. It is less than apparent how the public benefited from this sequence of events, except that people may have *felt* more secure about their food sources. Of course, perception is a powerful force in social and political life.

It is possible that increased cancer caused by pesticides has not yet appeared in epidemiological studies because the time between initiation and promotion of the disease has not yet passed. On the other hand, pesticides can also serve public health functions. When combating insect-vector diseases, such as yellow fever, bubonic plague, typhus, or malaria, pesticides can literally be lifesavers. In the years since DDT was banned, for example, malaria has reappeared as a major public health danger in Third World countries (Whelan 1994, pp. 100–104). So there are quite real public health benefits from pesticides as well as risks to weigh (Doll and Peto 1981, p. 1250).[31]

After the major epidemiological studies on cancer summarized in this chapter were completed, there was a controversy over the carcinogenic potential of Alar, a pesticide and growth-stimulation compound used by apple growers. In February 1989 the CBS television program *60 Minutes* broadcast a segment on a report by the Natural Resources Defense Council (NRDC), which charged that Alar was carcinogenic. Children were particularly at risk, according to the report, because of their small body weights. A controversy over Alar quickly exploded. Celebrities appeared in television ads scrubbing broccoli on behalf of a group called Mothers and Others for a Livable Planet. Surveys at the time showed that more than eight out of ten Americans viewed pesticide residues on food as a serious health problem. Apple sales dropped sharply. By June the chemical company making Alar canceled U.S. sales.[32] After these dramatic events, as scientific reports on the risk posed by Alar were reviewed, it was found that the NRDC report consistently made assumptions and used models that maximized the reported risk. Subsequent analysis by the California Department of Food and Agriculture found that the probable risk from Alar was 3.5 lifetime cases of cancer per *trillion* population (about twenty times the current population of the earth). The CDFA analysis indicates that the American population would have to increase a thousand times over before one cancer caused by Alar could be expected. By 1990 the anxiety over apples began to subside, but consumer concern over food safety remained high (See Rosen 1990; Marshall 1991). Indeed, the debate over Alar continued into 1996, when Elizabeth Whelan of the American Council of Science and Health criticized the *Columbia Journalism Review* for defending the NRDC's analysis of the dangers of Alar. Paul and Anne Ehrlich (1996) also continue to defend the report.[33]

Many foods, such as celery and pepper, contain natural carcinogens. In the aftermath of the Alar controversy, Bruce Ames and other scientists pointed out that public fears were being focused on cancer risks that are vanishingly small, while more potent but familiar carcinogens, such as secondhand cigarette smoke, are routinely accepted by the public (see Ames, Magaw, and Gold 1987; Zeckhauser and Viscusi 1990; and in a more popular form, Fumento 1993).

In 1996 a committee of the National Research Council (NRC), the research arm of the National Academy of Sciences, released a report on carcinogens in food that reflected the conclusions drawn from the epidemiological studies mentioned here. The report argues that the cancer threat in the human diet today comes primarily from diets too rich in calories and fats or alcohol and too low in vegetables (NRC 1996, pp. 336–337). The report describes many of the carcinogens and anticarcinogens, both natural and synthetic, found in foods and concludes that a balanced diet, high in fruits and vegetables, provides significant protection from toxicants in foods. It also concludes that the number of natural substances that are carcinogenic is probably at least as large as the number of synthetic substances that are.

In essence, this report says that much or all of the hoopla about food additives, traces of pesticides, and low concentrations of carcinogens in foods has been misleading. It maintains that the rodent tests at the time of the report were inadequate to understand the links, whatever they might be, between dietary chemicals and cancer. An earlier report by the NRC (1993) had been used to suggest that children might be somewhat more susceptible to cancer as the result of consuming pesticides on fruits and vegetables than are adults. Such anxieties would seem, however, to be unwarranted (Tufts University Diet and Nutrition Letter 1996b).[34]

A related controversy has swirled around the potential carcinogenic effects of dioxin. The herbicide Agent Orange, which contained dioxin, was used during the war in Vietnam to defoliate areas of jungle. Some American servicemen who were in Vietnam during the war later charged that their exposure to dioxin had damaged their health and caused birth defects among their children. Their suit against chemical companies that produced dioxin was resolved by a settlement in which the companies agreed to establish a large fund to compensate veterans who supposedly had been made ill or who had died as a result of Agent Orange exposure. Since exposure could never be proven, small payments of about $10,000 each were established for all completely disabled veterans and for survivors of veterans who had died.[35]

Meanwhile, the scientific case against dioxin as a carcinogen grew weaker. By 1991 government scientists were backing off claims of serious adverse health effects from dioxin. Though the process took longer than in the case of Alar, there were a number of parallels: alarms about cancer and other health effects, government action, reaction by affected individuals and businesses, and, over time, scientific findings that the original fears were greatly exaggerated (see Schneider 1992, 1994b; Harrison and Hoberg 1994).

In 1995 the EPA seemed to be about to change its mind again. Word was leaked to the press of an EPA study that once again implicated dioxin not only in causing cancer but also in harming fetuses whose mothers were exposed to significant levels of dioxin (Schneider 1994a; Lee 1994). But when the study was submitted to EPA's own advisory panel of scientists, the Science Advisory Board (SAB), its members literally tore the findings apart (Stone 1995). The Science Advisory Board report (SAB 1995) concluded that the only disease that had been convincingly linked to dioxin exposure is a skin disease called chloracne, that the parts of the report written by EPA staff have a tendency to overstate danger, and that the EPA method for estimating cancer risks exaggerates such risks. The EPA announced that it expected to respond to the SAB comments and produce a new version by the summer of 1996, but given that the first draft took four years and six million dollars to produce, that schedule seemed overly optimistic to observers. Thus far the skeptics have been proven right. The EPA has yet to respond.

Dioxin is intertwined with Agent Orange because the dioxin contaminant in Agent Orange was the risky part of the herbicide. While the EPA cannot convince the SAB of any problems caused by dioxin but chloracne, a committee of the Institute of Medicine (IOM) concluded in 1993 that there were other dangers, and the Department of Veteran's Affairs (DVA) began paying compensation. The IOM report and the DVA policy were enacted despite clear evidence that very few veterans of ground warfare had been exposed to

Agent Orange (Center for Disease Control 1988; Gough 1991) and that the Air Force personnel who handled and sprayed 90 percent of the Agent Orange used in Vietnam have morbidity and mortality rates that are essentially the same as those of Air Force personnel who were not exposed (United States Air Force 1995). More recently President Clinton has expanded compensation coverage so that payments are being made to veterans who have contracted prostate cancer and other diseases that, it is now argued, are associated with Agent Orange (Purdum 1996), despite the lack of evidence.

OTHER AND UNKNOWN CAUSES

In addition to the possible contributors to cancer that we have reviewed, concerns have been raised over such substances as aerosols and hair dyes. Similarly, after preliminary tests showed that the fire retardant "Tris" was carcinogenic, it was banned from use on children's sleepwear (Office of Technology Assessment 1981, pp. 94–95). In their review of current research, Doll and Peto stressed that although it seems unlikely that any major causes of cancer have missed detection, "there is too much ignorance for complacency to be justified" (Doll and Peto 1981, p. 1251). They simply entered a question mark under "unknown" causes of cancer. Higginson and Muir estimated 17 percent of male cancers to be either congenital (inherited) or unknown in origin. Wynder and Gori estimated that 16 percent of all cancers in men and 20 percent in women are of unknown origin.

Despite the continuing uncertainty, a few comments are possible. First, there seems to be a propensity for cancer in some families. Women whose close female relatives have developed breast cancer are more susceptible to it, as noted, and in some families the susceptibility extends to other forms of cancer. Is cancer sometimes an inherited disease? Close studies of "cancer families" show that it may be, but of course there are many complicating factors (Whelan 1994, pp. 192–5).

Second, it seems clear that at least a few forms of cancer are caused by infections. Probably the clearest example is Burkitt's lymphoma, in which a virus is implicated. As noted above, cancer of the cervix in women may well have origins in a communicable agent.

Certain bacteria and parasites may also induce forms of cancer, but the research on this possibility is very limited. Doll and Peto make a "very uncertain" estimate that 10 percent of all cancers are caused by infection. This figure is consistent with the range of unknown causes included by Higginson and Muir, and Wynder and Gori (Doll and Peto 1981, p. 1255).

Third, because of the synergistic aspects of carcinogenesis, there may be complex, interdependent chains of cancer causation that scientists do not yet understand. This does not mean "everything causes cancer," but there are limits to scientific understanding of carcinogenesis at this time (see Office of Technology Assessment 1981, p. 66; Malkin and Portwine 1994; Van Loon et al. 1995; Whelan 1994, p. 336).

Counting and Knowing: The Methodology of Cancer Epidemiology

Despite the continuing uncertainty, it is valuable to understand how we know what we *do* know; that is, the methodology of environmental carcinogenesis. Indeed, much of the debate over environmental cancer revolves around controversies about the methods of collecting and analyzing data about the incidence and nature of environmental cancer.

To follow the methodological debate, it is necessary to know that (1) the longevity enjoyed by Americans today makes cancer rates look higher than they were when the average life span was shorter; (2) there are significant differences between the data on cancer now being gathered and those collected in past decades—that is, greater access to better health care with improved diagnostic techniques makes things look worse because more cancers are identified; and (3) cancer data should be viewed site by site, rather than in the aggregate. Because of the importance of these factors, we shall consider them in some detail.

LONGEVITY

The longer people live, the more opportunity they have to encounter all types of disasters, but especially cancer. Cancer is disproportionately a disease of old age. One reason more people are dying of cancer is that they are not dying of other things first.

Infant mortality in the United States is substantially lower than it was at the turn of the century. Influenza, tuberculosis, poliomyelitis, and dysentery are now responsible for only a tiny fraction of the suffering they once caused. Public health measures, including food inspection, proper sewage treatment, and a clean water supply, are better than ever before. The main reason for rapid population growth in the Third World has not been an increase in birth rates as much as vast improvements in diet, sanitation, and medical care. In the United States, in 1900 fewer than 4 percent of all deaths were attributed to cancer. By 1935 cancer was said to account for 9 percent of all deaths among men. By 1975 the disease claimed fully 19 percent of those who died (see Richards 1972, p. 121; Blumenthal 1978, pp. 12–13; Doll and Peto 1981, p. 1208).

Herman T. Blumenthal reports that about 30 percent of the increase in cancer deaths since 1900 can be accounted for by increased life expectancy. To put it another way, at age twenty-five the chances that a person will develop cancer in the next five years of his or her life is only one in seven hundred. By age sixty-five, however, the odds are one in forty-four that cancer will develop before age seventy (Blumenthal 1978, p. 13). Blumenthal suggests that the process of aging itself may contribute to carcinogenesis. That is, the cells in a seventy-year-old's body are apparently more disposed to turn cancerous than those in the body of a twenty-five-year-old. But one need not accept this hypothesis to see that the "aging" of America has had the effect of increasing the number of deaths due to cancer.

For methodological purposes, the aging of the American population means that cancer rates must be adjusted for age. Even if everything else were equal, the raw number of cancer cases and deaths would be greater today than forty years ago because so many more elderly people are now in the population. Therefore serious studies use "age-adjusted" cancer rates.[36]

BETTER DATA/WORSE APPEARANCES

Since the turn of the twentieth century, vital statistics on the cause of death have been collected by the federal government. Beginning with the First National Cancer Survey (FNCS) in the 1930s, efforts were made to collect national information on the incidence of cancer. Incidence measures the number of cases of cancer, whether the patient dies or recovers. Epidemiologists use incidence data to understand the course of the disease, but it is much less reliable than mortality data. The FNCS was followed in the late 1940s by the Second National Cancer Survey, and in the late 1960s and early 1970s by the Third National Cancer Survey (SNCS and TNCS). Then, beginning in 1973, the National Cancer Institute started the Surveillance, Epidemiology, and End Results (SEER) program to monitor cancer incidence and mortality each year.[37]

As time has passed, the quality of the data collected in these programs has improved, meaning that many cases of cancer that would have been missed in the First or Second National Cancer Surveys are now reported. The Office of Technology Assessment reports numerous reasons for the errors in the old data. Among the most important are:

(1) Improper diagnosis—not only has cancer become more common in recent years, it has become more acceptable to discuss cancer. Doctors and other health workers have become used to considering and listing it as a cause of death.
(2) Improvements in ascertainment—for reasons parallel to earlier diagnostic underreporting, fewer cancers are now missed altogether. In the past it seems likely that a notable number of cancer deaths were reported as deaths from "natural causes."
(3) Changes in the definition of cancer—as categories of cancer are added, the number of reported cases of the disease increase, even though there is no real increase in the number of people becoming ill.
(4) Increased access to medical care for the old—the availability of Medicare means that many elderly patients who could not afford to see a physician during the time of the First and Second National Cancer Surveys were in a position to do so by the time of the Third and have been able to do so since the SEER program began.
(5) Increased access to medical care for the poor—Medicaid has had much the same impact among poor people as Medicare has

among the elderly. This may in part account for very sharp reported increases in cancer rates among African-Americans, although cancer incidence was still underreported in minority populations in at least the first and second surveys (Office of Technology Assessment 1981, p. 40).

SITE-BY-SITE ANALYSIS

It is a mistake to consider cancer data in the aggregate. Cancer is not one disease and does not have one cause. In the United States stomach cancer, for example, is down and lung cancer is up substantially. Because of the categories we use to count cancers, skin cancer seems to have increased greatly in the past two decades—but that increase is largely explained as a function of AIDS.[38] The incidence of many other cancers has remained essentially stable. We have already considered the significance of such differences for the debate over cancer causation. These considerations of cancer site, improved methods of medical treatment, and increasing longevity prove useful in sorting out three major disputes over cancer that have revolved around methodology: first, the debate over the estimate that 90 percent of all cancers are "environmental" in origin; second, the disputed estimate from Joseph Califano (and the OSHA paper) that up to 40 percent of all cancers are or will be caused by occupational exposures to carcinogens (discussed above); and third, the claim that there is or will soon be a cancer "epidemic" in the United States.

The notion that 90 percent of all cancers might have environmental origins can be traced back to speculation by Dr. John Higginson at an international conference of cancer researchers. By now it should be clear that Higginson's estimate has frequently been misappropriated. Following the epidemiological convention, he was defining "environmental" to include smoking, diet, sex, and other personal habits, as well as causes that are more typically assigned to the external environment in ordinary language, such as occupational hazards and pollution. Higginson denounced the misuse of his work by sensationalist writers and publicists. He explained that the "environment is what surrounds people and impinges on them. The air you breathe, the culture you live in, the agricultural habits of your community, the social cultural habits, the social pressures, the physical chemicals . . . and so on."[39]

We have already discussed the methodological problems in the OSHA paper, which had the effect of exaggerating real but limited dangers from occupational exposures to carcinogens. It is enough to say here that when those methods were exposed, the resulting estimates were rejected by epidemiologists and other serious scientists as "clearly inflated," to borrow a phrase from the pages of *Science*. The same article continued, "one might find that the risk of the driver dying in an automobile crash is one in ten if the automobile is consistently driven at speeds in excess of 120 miles per hour, and that there are currently 100 million automobiles on American highways. Using the logic of the government report, one would conclude that there will be 10 million excess deaths as a result of driving at high speeds" (quoted in Proctor 1995, p. 65; see also pp. 64–71).

The last of the methodological problems concerns the claim that the United States is now, or is about to be, overwhelmed by an "epidemic" of cancer. This claim falls into the methodological discussion because it unravels when the data are disaggregated into site-by-site trends (*SEER Cancer Statistics Review, 1973–1991: Tables and Graphs* 1994). There is no question that lung and other respiratory cancers, along with cancers related to AIDS, had been rapidly increasing. Were these the only cancers, the disease might almost be growing at epidemic rates. Moreover, past trends suggest that the number of women who will develop lung and other respiratory cancers will increase substantially because the number of female smokers increased significantly in past decades.

The increases (in some cases only apparent) in lung, breast, and prostate cancer, and quite recently cancers associated with AIDS (particularly Kaposi's sarcoma and non-Hodgkin's lymphoma) have been large enough that overall cancer incidence rates were going up in the United States between 1973 and 1990. Cancers at other sites, however, were either stable (if minor variations are dismissed owing to the unreliable nature of older data on the disease) or were declining slightly (see Devesa et al. 1995).

Beyond the large increases in respiratory cancers, site-by-site analysis showed a basically stable (or slightly declining) pattern.

Stomach cancer was declining in the United States, perhaps because of improved food storage and consumption of reduced amounts of contaminated food. Intestinal cancer, including colon cancer, had been largely stable during the period for which data were available. Liver cancer is a rare disease and appears to be stable. Skin cancer is increasing as Americans expose themselves to the sun more frequently now than in the past and because Kaposi's sarcoma, which is associated with AIDS, happens to be categorized as a form of skin cancer. Fortunately, most skin cancers are relatively easy to treat. Melanoma, the often fatal variety of skin cancer, is increasing but still rare.

Breast cancer had increased in the 1970s and 1980s in the United States. Bladder cancer rates were largely stable, which is somewhat surprising since smoking contributes to bladder cancer, but it is possible that improved diagnosis may have more accurately defined what once would have been called bladder cancer as bladder metastases of cancers that originated in other sites. Mortality from uterine cancer is declining, in part because of early detection.[40]

Improvements in medical and scientific technology have contributed to public anxiety about cancer in other ways. Science now has a greatly enhanced capacity to ferret out trace amounts of impurities and suspected carcinogens using new equipment, such as electron microscopes, and new chemical tests. These can detect a single part per trillion—the same relation as a penny to ten billion dollars. It is reasonable to pay attention to potential risks associated with the accumulation of a part per billion here and a part per billion there, but the body does have defenses, and the research to date has not uncovered any carcinogen that is likely to act at these concentrations.[41]

The Cancer Issue Network and Public Policy

Whatever the scientific evidence, human cancer is also a politicized disease. To understand the politics of cancer requires an excursion into the cancer "issue network."[42] Traditionally, understanding any specific public policy in the American political system has been simplified by knowing that in most cases three main participants establish and maintain policy: interest groups with a material stake in the outcome of the policy choices made; the bureaucracy responsible for carrying out policy; and the congressional committees (including the appropriations subcommittees) with jurisdiction over the matter in question. These relationships were called "policy subgovernments" or, informally, at Washington cocktail parties, the "iron triangles," because of their legendary strength and resistance to outside influences.[43]

In the years following World War II, when these little-discussed arrangements were very strong indeed, health had not yet arrived as a major responsibility of the national government. But we can observe the iron triangle at work in the making of science policy. Thus, Don K. Price could write of a "scientific establishment" in the old sense of the word, that is, an institution for which taxes are collected and funds spent almost on faith and "under concordats which protect the autonomy, if not the cloistered calm, of the laboratory."[44]

Like so much else that was straightforward about the old system, the unchallenged position of scientists as masters of their own political house has largely unraveled in the new political system in Washington. That is not to say that scientific interest groups have become powerless. Nor would anyone familiar with science politics deny the influence of the relevant congressional committees or the many agencies involved, including the National Science Foundation, the National Institutes of Health, the Environmental Protection Agency, or the Food and Drug Administration. But a wide range of groups and individuals who were very much on the outside looking in at the scientific establishment are now actively hauling and pulling at the reins of science policy. Cancer policy has become prominent enough to have its own specialized "issue network," including scientists, bureaucrats, members of Congress and their staffs, White House staffers, journalists, lawyers, "public interest" group lobbyists on both the left and the right, union and business representatives, state and local government leaders, and others.[45] Among the newest activist groups is the National Breast Cancer Coalition, an organization that sees its function as placing women who have come to understand issues of breast cancer in a position to influence research and treatment policies (*Science* 1995b).

An overview of the relevant major laws affecting national cancer policy, the dominant interpretations of those laws, the major institutional "players" in the cancer issue network, and the context of the approach to risk taken by this network will help us understand the new politics of cancer.

THE LEGAL FRAMEWORK

There are at least ten federal laws that restrict or regulate carcinogens. Probably the most important was the now repealed Delaney clause of the Food, Drug, and Cosmetic Act, which is administered by the FDA. In addition, a number of laws administered by the EPA aim at environmental cancer; among these are the Clean Air Act, the Water Pollution Control Act, the Federal Insecticide, Fungicide, and Rodenticide Act (FIFRA), the Federal Environmental Pesticide Control Act, the Resource Conservation and Recovery Act (RCRA), the Safe Drinking Water Act, and the Toxic Substances Control Act. The Consumer Product Safety Commission can also cite two laws that empower it to act against cancer risks to consumers: the Federal Hazardous Substances Act and the Consumer Product Safety Act. Finally, the Occupational Safety and Health Act can be read as including some of the most explicit language regulating potential carcinogens in the workplace; OSHA is clearly empowered to act against cancer risks in the workplace (Office of Technology Assessment 1981, pp. 176–181). So there is no lack of authority to act.

In these pieces of legislation and judicial decisions interpreting the laws, three standards for assessing and managing risk are reflected. The first of these, in the Delaney clause and the Clean Air Act, is the absolute standard, which holds that *no* risk from cancer is acceptable. The second, reflected in FIFRA, the Safe Drinking Water Act and the Toxic Substances Control Act, is the balancing standard, which attempts to weigh benefits against risks. For example, the very small possibility that chlorine in drinking water may be carcinogenic must be balanced against the enormous public health benefits that accrue from chlorination. Third, some clauses of some laws reflect a technology standard for controlling risk. The most celebrated of these are the "technology-forcing" (also known as "agency-forcing") standards (or clauses) of the Clean Air Act and the Water Pollution Control Act, which require installation of the "best available technology" (BAT) or the "best practical technology" (BPT).[46]

Because the absolute standard of the Delaney clause conferred the greatest power to the agencies, and because it was the hardest to satisfy, it deserves some attention even though it has been repealed. In 1958 the Delaney clause reflected the existing understanding of scientists regarding cancer and its causes. (It also made some assumptions about the capacity of scientific devices to detect small quantities of potential carcinogens in food). Still, the passage of the Delaney clause also showed the influence that scientists allied with members of Congress can have. Wilhelm Hueper provided evidence for the Select Committee Investigating the Use of Chemicals in Food and Cosmetics. Later, Hueper presented to Congress the recommendations of a 1956 science conference in Rome on environmental carcinogenesis (see Agran 1977, pp. 180–185 for details). There was nothing unusual or sinister in this; such arrangements were and are routine all over official Washington. But the linkage illustrates the importance of experts who act as interpreters of the scientific state of knowledge to policymakers. Law in the regulation of science is partly based on research findings, but it is more substantially the result of the interpretation of basic research by experts trusted in the issue network.

INTERPRETING THE LAW

In practice, the operation of the cancer prevention objective embodied in legislation has depended largely upon animal tests (Cushman 1996a), though a recent statement by the EPA will bring about changes. Such tests have typically used mice or rats that are fed doses of a chemical or food additive at varying levels of potency, sometimes at concentrations so intense that the chemical being tested becomes toxic, even though it is benign at lower concentrations (Ottoboni 1984).[47] After the deaths of the animals, their organs are bioassayed, that is, examined microscopically for tumors.

There are problems with animal tests, to be sure, but short of simply introducing substances into the human environment, they have been considered essential to predicting the potential for carcinogenesis in humans (see, for example, Rosen 1990). But a major problem with animal tests is that the results must be interpreted by people. Regulatory agency personnel tend to have, or to develop, an approach that favors the most "prudent"—that is, the most risk-aversive—alternative available (see Zeckhauser and Viscusi 1990, pp. 560–561). If this approach really protects the public from carcinogens, it is admirable, but doubts have been raised by a number of distinguished cancer researchers, such as Bruce Ames and Gio B. Gori (see Ames, Magaw, and Gold 1987; Gori 1980), who seem finally to have convinced the EPA of the validity of their critique.[48]

But even if scientists at the FDA and the EPA do take a different view of their responsibilities, interpretation of environmental and human dangers will not necessarily change much, because many other actors now play a central role in determining what procedures will be followed and, more important, how results are to be interpreted. In the 1950s federal regulatory agencies were subject to many checks and balances in the political system, but the presumption of the statutes creating them was that they would exercise substantial independent discretionary authority. Thus, FDA and other government scientists enjoyed a relatively high level of professional autonomy when they made judgments.

That state of affairs is considerably less true today, in part because laws were rewritten by Congress to give outside groups a basis for challenging the judgments of agency scientists and administrators, and in part because broad judicial doctrines regarding standing to sue expanded considerably. The Clean Air Act, for example, includes provisions written by Congress to allow judicial second-guessing of the EPA by environmentalists.[49] But even without these changes, judicial acceptance of challenges to regulatory decisions means that the most sensitive decisions will almost certainly be reviewed in court. Judges have been increasingly willing to referee scientific controversies (see Lunch 1987, chap. 5; Wilson 1989, chap. 15).

THE MAJOR PLAYERS

Significant changes in employing a very high if not an absolute standard for controlling risk seem unlikely in the foreseeable future. Administrators at FDA, EPA, OSHA, CPSC, and similar agencies have a natural interest in maintaining the conditions for a maximum level of regulation. Moreover, there are important constituencies in the United States, including most environmental groups, that are committed to maintaining if not raising existing standards in assessing risk. In an era of divided government, at least when the Democrats have controlled the House and Senate, congressional committees in the issue network have been likely to share that view.

A number of congressional committees and subcommittees have partial jurisdiction over regulation of cancer risk, but during most of the past twenty years, the Health and the Environment Subcommittee of the House Commerce Committee, chaired from 1981 to 1994 by Representative Henry Waxman of California, was dominant. Waxman is a Democrat from Los Angeles whose constituents include a large number of elderly people, environmentalists, and other citizens who routinely live with air pollution, so it is not surprising that he was a leading opponent of the Reagan and Bush administrations efforts to change environmental laws, including cancer policies. Because the environmental issue network includes cancer policy, Waxman was a pivotal figure for a decade and a half.[50]

Since 1995, however, Republicans have been in control of Congress and its committees. The chair of the Health and Environment Subcommittee is Michael Bilirakis, a Florida Republican from a district near Saint Petersburg with many elderly constituents.[51] The new Republican House leadership, however, has given greater authority to full committee chairs than the Democrats did after the mid-1970s. Thus, substantial authority over cancer policy (and environmental policy more broadly) may be exercised by Thomas Bliley, the new chairman of the House Commerce Committee, to which the environmental subcommittee is attached.[52]

Bliley represents a district in and around Richmond, Virginia. He has a record of opposition to many environmental laws and the existing level of regulation of drugs. He may well have influence in cancer politics in a number of ways. First, as a member of the House from Virginia, where the tobacco industry is a major employer and is more popular than in most states, Bliley has attempted to block further restrictions on tobacco sales and use, if not roll back those restrictions already in place. Second, he is seeking to speed up regulatory approval of drugs by the FDA. Third, he has been visible as a sponsor of proposals to restrict the Clean Water Act, one of the key pieces of legislation that contributes to national cancer policy.

The Delaney clause was a specific target for some conservatives and Republicans. Early in 1995 Senator Charles Grassley, a Republican from Iowa, introduced an amendment to the Republican regulatory relief bill in the Senate that would notably weaken the Delaney clause; in the House, a bill to explicitly repeal the clause was also introduced. Legislation achieving this end was passed by Congress and signed by President Clinton, though other rules replaced the clause, whose effect on regulation is difficult to assess as yet (Freedman 1998). As noted earlier, the legislation was able to pass Congress and be signed into law because of the widespread agreement about it in the scientific community which allowed Bliley and Democratic representative John Dingell to work out a compromise.

As this partial list of proposals and goals suggests, the new Republican majority in Congress had an ambitious—or destructive, depending on one's point of view—set of goals for limiting environmental and health-and-safety laws that would have substantial impact on cancer policy.[53] Their efforts to change policy in the 104th Congress largely failed, and the 1996 election weakened their effort. Indeed, such is the popularity of environmental regulation that their attempt to produce change in this area probably contributed to a reduction in their congressional majority. This is especially true of their efforts to pass legislation that would mandate a general weighing of risks and benefits in making decisions about environmental issues.

The ineptitude of the Republicans in many areas helps explain their inability to move faster. They have been faced, however, with a skeptical press and a situation in which many of their experienced staff members were less than fully sympathetic to their efforts. Indeed, by 1997 they were in some instances, to paraphrase a headline in the *New York Times,* courting the greens (Berke 1997).

In retrospect, the political heavy-handedness of the Reagan and Bush administrations regarding environmental legislation was often self-defeating. In the early 1980s, for example, the Office of the Science Advisor in the White House started a review of cancer policies and was reportedly planning changes that would have been unacceptable to the relevant congressional committees. The effort was aborted after scandals at the EPA caused officials to view all environmentally related issues as too politically damaging to pursue further (see Wines 1983, p. 1268).

Waxman and other Democrats have regularly been able to point to national public-opinion polling in support of their position.[54] But beyond the specific members of Congress in powerful positions at any given time, it is likely that environmentalists, labor, and their political allies will remain on one side, with business and conservatives on the other. Cancer policy, perceived as a subset of environmental policy, is likely to continue to respond to pressures similar to those that created the Alar controversy as long as policy is framed in an adversarial, labor-versus-management, environmentalist-versus-Philistine context.

Among the environmental groups, the Sierra Club, the Natural Resources Defense Council (NRDC), the Environmental Defense Fund, the Center for Science and the Public Interest, Friends of the Earth, Greenpeace, the Audubon Society, the National Wildlife Federation, and the Wilderness Society are only some of the more prominent organizations that are prepared to do battle on this topic. Members of Congress must respect these organizations for their capacity to mobilize campaign volunteers, influence public opinion, and raise campaign funds.

On the other side, the drug and chemical industries have a large stake in cancer policies. Whatever position these businesses take, the public will view them as hopelessly tainted, both by their own misdeeds of the past and by the material interest they have in cancer policy, as even private polling by industry has found.[55] The charge of profiteering from the misery of others is taken very seriously. For example, there is deep distrust of business among environmentalists and many members of Congress, as the political scientist Steven Kelman found when he investigated the idea of allowing industries to "buy" pollution rights with higher tax payments to give them an economic incentive to clean up wastes. There was a strong ideological dimension to opposition to the idea. "I think there is something anomalous about taxing a poison. A tax approach appears to sanction the poisoning," said one environmental activist. Another declared,

"A crime against nature is a crime against society . . . if my son dies of cancer, I want to blame somebody, and I want somebody to be accountable."[56]

The adversarial perception of cancer policy flows not only from past experience but also from contemporary perception. Press treatment of the issue of environmental carcinogenesis has sometimes been alarmist, as the coverage of both the OSHA paper and Alar illustrate. To be fair, few reporters are trained in science, even if they cover science. But research done at the National Cancer Institute indicated that between 1977 and 1980, years when cancer was a hot topic for the press, the most frequently mentioned causes of cancer in the fifty largest newspapers in the nation were pollution, occupational exposure, chemicals, and food additives (research cited by Boffey 1984). At roughly the same time some of the leading medical textbooks on cancer paid little or no attention to these variables in discussing the etiology of cancer (for example, DeVita, Helman, and Rosenberg 1985). Does such coverage continue now? What are the views of those reporting on such topics? How are popular attitudes toward cancer policy shaped? These are the questions to which we now turn.

4
The Experts versus the Activists

WHO KNOWS WHAT CAUSES CANCER? As we have seen, this is one of the most controversial, contentious, and politically charged scientific questions of recent decades. But another question may shed light on this debate: Who *cares* what causes cancer? Some of the most powerful forces in American society are involved in the continuing struggle to translate scientific information into public policy on environmental cancer.

Who has an interest in this knowledge? Let us count the players. There are business interests whose products may be stigmatized or banned outright, or whose health and safety practices may invite costly litigation. There are trial lawyers whose interest lies in pursuing just such litigation. There are labor unions vigorously protecting their members against long-term occupational exposure to substances that are more feared than understood.

Of course, the contest extends to the public sector, where personal and institutional interests seem just as ubiquitous, if less clearly identifiable with financial self-interest. There are politicians seeking the moral high ground in the quest to "protect the public health," "save American jobs," or "punish the evildoers." Less visible but equally important are the bureaucratic struggles, both between and within government agencies like EPA, OSHA, NIH, and FDA, each looking to take the lead in promulgating cancer policy.

Finally, there are environmental and public health watchdog groups, which claim to represent the public interest against self-seeking executives, politicians, and bureaucrats. These nonprofit research and advocacy groups use research, publicity, litigation, and grass-roots mobilization to put across their perspectives on health threats in the environment.

But what about the experts—the independent cancer researchers whose careers revolve around the study this disease (or diseases) and whose goal is to conquer it? They do not seem to have a regular role in the public debate, either as active participants or as *dei ex machina* who descend from ivory towers to sort out controversies involving the common weal. Indeed, at times they seem to be speaking a different language from the other players in this drama, as we learned by contrasting the scientific and political usages of the very term *environmental cancer*.

Particular individuals with scientific credentials have often been influential in the debate over cancer policy. These emissaries from the scientific community to the public typically include the heads of a few government agencies such as the National Cancer Institute (for example, Samuel Broder), researchers who gain renown for their highly publicized discoveries and awards (David Baltimore), and activists who represent public interest groups (Sidney Wolfe).

Yet we cannot assume that even the most eminent or recognizable scientists speak for their peers in their public pronouncements. The careers of researchers as illustrious as Linus Pauling and William Shockley illustrate the dangers of doing so. Even the public panels and commissions that are sometimes convened may have their membership skewed by arbitrary selection procedures and their reports colored by group dynamics or other nonsubstantive factors.

We have argued that the views of the expert *community* have had little impact on policymaking in this field, in contrast to the views of individual scientists or small groups whose influence derives from their institutional positions or their political activism. The evidence presented in chapter 3 reflects our own reading of the literature, especially the estimates of cancer causation produced by several prominent epidemiologists during the late 1970s and early 1980s. But these estimates reflect a knowledge base that is now fifteen or twenty years old. How can we know whether they remain relevant to such a diverse and rapidly changing field of inquiry? In any case, critics might argue that our selection and interpretation of these data are skewed by our own biases. How can we demonstrate that our version of the expert consensus on cancer causation is a valid reflection of the actual expert community?

Moreover, if the relevant experts are largely ignored in this process, where does the information base for the policy debate come from? How are public misconceptions created and spread? What alternative expertise or perspective sustains them against the mainstream of scientific thinking? We will begin to answer these questions by presenting a systematic portrait of scientific opinion and counterposing it against the comparable opinions and broader outlooks of leading environmental activists.

In the public debate over environmental cancer, the "green lobby" has replaced academic scientists as the major source of information about the environment and, for our purposes, environmental cancer, for the media and the general public including high-status public officials.[1] But to understand how one perspective has come to supersede the other among the general public, it will be necessary to shift the focus from the creation of knowledge to its dissemination. This is the task to which we will now turn by tracing the media's treatment of these ideas over the past quarter-century.

Polling the Experts

In most public policy arguments, each side relies on its own favored "experts," some of whom exhibit more skill with sound bites than with scientific evidence. The cancer issue is no exception—except that, in this case, the most relevant independent body of knowledge and expertise on the issue has *not* been heard much in the public discourse. Cancer scientists in the United States have hardly been silent on these matters, as a wealth of scholarly articles in research journals attest. Until now, however, their collective voice has been heard by few outside the scientific community.

In order to introduce this voice into the public debate, it is vital that it be truly representative of the expert community, rather than another self-selected or self-interested subset of that group. To achieve this end we applied the techniques of survey research based on random sampling. Instead of presenting the views of "our" favored experts, we sought to reproduce the actual spectrum of opinion that exists among scientific researchers in this field.

For expert opinion on the science of cancer causation we turned to the membership of the American Association for Cancer Research (AACR). This organization is made up of research scientists who have established a track record of peer-reviewed publications on the causes, treatment, and prevention of cancer. It is the most broadly based and intellectually prestigious professional association of cancer researchers in the United States.

To secure a representative portrait of cancer expertise, we commissioned the Roper Center for Public Opinion Research to survey a random sample of active AACR members who listed either carcinogenesis or epidemiology as their field of specialization. The Roper Center is an academically affiliated nonprofit public opinion research facility that has extensive experience in surveying expert groups. This facility also maintains the nation's largest archive of survey data, which permits ready comparisons with independently collected data from national samples.

Working in cooperation with the center's senior researchers, we designed a questionnaire that addressed the major issues in cancer causation and policy described in chapter 3. The questionnaire consisted of ninety items concerning the contribution of numerous environmental factors to human cancer rates, scientific controversies related to cancer research, and assessments of various sources of information on environmental cancer, including the media. It also asked for information on the professional and social backgrounds of those surveyed.

After drawing an appropriate sample from the AACR's 1992 membership list, the Roper Center administered the survey to 401 cancer researchers in telephone interviews conducted during January and February 1993. Sampling error for a survey of this size is +/-5 percent. As an indication of their success in reaching the appropriate expert population, 92 percent of those interviewed said that they were currently involved in research on the causes or prevention of cancer, and a majority had published forty or more articles on these topics in peer-reviewed journals. In addition, 90 percent described their principal current position as that of faculty member or researcher.

The largest portion of the survey asked scientists to rate both specific substances and general aspects of the environment in terms of their relative contribution to cancer rates in the United States. Two separate lists of ratings were used in order to avoid inappropriate comparisons, such as the relative contributions of pesticides in general and DDT in particular.

Because of its special status as a widely recognized major carcinogen in a variety of forms, aspects of tobacco use were included in both lists. In other sections, the cancer scientists were also asked to rate the media's treatment of some of these risk factors, as well as the overall reliability of several information sources.

ENVIRONMENTAL SOURCES OF CANCER

The cancer experts were asked to evaluate thirteen aspects of the environment in terms of their contribution to human cancer rates in the United States. They assigned each aspect a number on a scale of 0 to 10, where 0 indicates that the environmental agent in question makes no contribution to cancer rates, and 10 indicates that it makes a very important contribution. (The midpoint on this scale is a rating of 5, whereas the midpoint of the more traditional one-to-ten scale is actually 5.5.)

In presenting the experts' ratings on this dimension, table 4.1 ranks each environmental agent according to the average (mean) score that it received. Also listed is the proportion of scientists who rated each agent as a "major" cause (7 to 10 on the scale), a "moderate" cause (4 to 6) or a "minor" cause (0 to 3). The category containing a plurality of responses is indicated in boldface.[2]

Tobacco smoke stood apart as the consensual choice for America's leading carcinogen. It received an average rating of 9.21 on the scale, amid virtually unanimous agreement (96 percent) that it is a major cause of cancer. This emphasis on tobacco will come as no

Table 4.1
Environmental Factors as Possible Causes of Cancer: Scientist Survey Ratings on 0–10 Scale

	Mean	Level of Concern (%)[a] About Aspects			
	Score	Major	Moderate	Minor	DK
Tobacco smoke	9.21	**96%**[b]	3%	1%	0%
Diet	6.38	**52**	34	11	4
Sunlight	6.33	**54**	32	13	2
Environmental smoke	5.74	**42**	35	22	1
Chemicals in workplace	5.44	**37**	27	25	3
Sexually transmitted disease	4.74	28	**34**	**34**	4
Pesticides and herbicides	4.72	26	**36**	34	4
Air and water pollution	4.70	26	**37**	34	3
Infectious diseases	3.96	14	36	**43**	6
Drugs (medical)	3.64	13	27	**54**	6
Food additives and preservatives	3.27	10	29	**57**	4
Chemicals in the home	3.05	7	27	**62**	4
Radiation (medical/dental)	3.00	10	22	**67**	1

[a] *Major = rating of 7 to 10; Moderate = rating of 4 to 6; Minor = rating of 0 to 3; DK = don't know*
[b] *Boldface = modal rating (plurality)*

surprise to even the most casual observers of the cancer debate. Indeed, it serves as a kind of baseline against which the contribution of other carcinogens may be gauged.

The next items on the list, however, might not be so easily predicted by consumers who worry about the pesticides they ingest, the pollutants they breathe, or the household chemicals they use. Instead, the second tier of scientific concern was reserved for dietary factors and sunlight. Each was named as a major contributor to cancer rates by a slight majority (54 percent for sunlight and 52 percent for diet). Their mean scores of slightly more than 6 placed them a long way down the scale from tobacco but well above the midpoint of the mortality spectrum.

Moreover, only two other factors were named as major cancer agents by even a plurality of scientists: 42 percent cited the effects of "environmental" tobacco smoke (secondary exposure to smoke in public places), and 37 percent cited chemicals in the workplace. These were also the only other factors to place above the midpoint on the scale, with mean ratings of 5.74 for environmental smoke and 5.44 for workplace chemicals.

Clustered together about one point lower on the scale were sexually transmitted diseases, pesticides and herbicides, and air and water pollution. Each was rated as a major cancer contributor by just over one in four scientists, and as a minor contributor by about one in three. Five remaining aspects of the environment were placed at the bottom rung of the ladder—infectious diseases, drugs used in medical treatment, food additives and preservatives, chemicals in the home, and radiation treatment. With the exception of infectious diseases, a majority of experts rated these agents as only minor contributors to cancer rates, and very few saw any of them as major contributors.

This pattern of results reflects priorities similar to those emphasized more than a decade earlier in the epidemiological studies described in chapter 3. Thus, diet and sunlight ranked well above all other factors aside from tobacco. The similarities to the earlier estimates are particularly strong with regard to the controversy over man-made substances.

Such high-profile risk factors as pesticides and pollution were ranked beneath sexually transmitted diseases as contributors to cancer rates, with mean ratings below the midpoint of 5 on the 0-to-10 scale. Similarly, food additives and household chemicals ranked below all factors except medical radiation, with majorities rating each as a minor carcinogen.

These rankings do not, however, replicate the earlier estimates in all instances. For example, our respondents displayed less concern about infection than did Doll and Peto (1981). Further, all three sets of estimates described in chapter 3 rated diet or "life-style" as an even greater cancer threat than tobacco. Such differences probably reflect the vastly expanded knowledge base of current researchers and variations in the way the question was addressed. In general, though, the survey data provide massive confirmation of the earlier estimates. Further evidence as to the stability of the Roper findings over time is presented below.

RISKS OF SPECIFIC SUBSTANCES

The Roper survey was designed to explore not only the cancer threats that scientists attribute to broad aspects of the environment but also specific substances that periodically generate public outcries. Therefore, we asked respondents to rate seventeen substances that have occasioned public controversy as suspected causes of cancer. The results appear in table 4.2.

Once again, cancer scientists placed tobacco in a league of its own among cancer agents. Indeed, only one in twenty researchers rated smoking as less than a major carcinogen. More notable, perhaps, was the high rating given to chewing tobacco, which two-thirds of these scientists also regarded as a major contributor to cancer rates. That rating placed this form of tobacco well ahead of asbestos, the only other substance named as a major cancer agent by a majority (56 percent) of experts.[3]

Secondhand tobacco smoke was the only other substance deemed a major cause of cancer by a plurality of cancer experts (46 percent). Thus, various forms of tobacco accounted for three of the top four substances on this rating list of dubious distinction. No other substance was rated as a major contributor to cancer rates by more than one-third of those in the sample. Only "fat in the diet" (5.39), of any of the thirteen other substances, received a mean rating above the 5 midpoint on the scale.

A cluster of five substances elicited a lower level of concern, with mean ratings ranging from roughly 4.6 to 5.4. In descending order, these included fat in the diet, the natural chemical aflatoxin, low fiber in the diet, dioxin, and alcohol. About one in three scientists rated high fat diets, aflatoxin, and dioxin as major contributors to cancer rates, compared to one in four who expressed as much concern about low dietary fiber. But dioxin's average rating dropped below that of both dietary factors, owing to the relatively large proportion of experts (40 percent) who rated it as only a minor cause

Table 4.2
Substances as Possible Causes of Cancer:
Scientist Survey Ratings on 0–10 Scale

		Level of Concern (%)[a] About Substances			
	Mean Score	Major	Moderate	Minor	DK
Smoking tobacco	9.19	**95%**[b]	4%	1%	0%
Chewing tobacco	7.34	**66**	19	14	1
Asbestos	6.49	**56**	22	20	2
Second-hand smoke	5.88	**46**	33	20	1
Fat in diet	5.39	33	**44**	21	3
Aflatoxin	4.85	34	**48**	13	4
Low fiber in diet	4.83	26	**41**	31	2
Dioxin	4.74	33	22	**40**	5
Alcohol	4.59	17	33	**48**	2
EDB	4.22	17	16	30	**47**
Radon	4.00	18	28	**49**	5
Hormones[c]	3.99	14	37	**45**	4
DDT	3.83	21	20	**52**	7
Nuclear Power	2.46	7	16	**73**	4
Alar	2.18	6	12	**64**	18
Saccharin	1.64	3	9	**75**	3
Other Sweeteners	1.19	1	6	**83**	10

[a] *Major = rating of 7 to 10; Moderate = rating of 4 to 6; Minor = rating of 0 to 3; DK = don't know*
[b] *Boldface = modal rating (plurality)*
[c] *I.e., as used in treatments*

of cancer. Indeed the mean score for dioxin was below that of dietary factors in general and aflatoxin in particular.

Beginning with dioxin, the remaining ten substances were all rated as minor causes of cancer by a plurality of experts who expressed an opinion. Four of these were clustered one increment lower on the rating scale—EDB, radon, hormones used as drugs (as in birth control pills and estrogen treatments), and DDT.

EDB was notable for the high degree of uncertainty expressed by those interviewed. A plurality of 47 percent was unable to rate its contribution to cancer rates, more than twice the proportion elicited by any other substance or aspect of the environment. (This did not affect its ranking, since the mean scores exclude respondents who were unable to rate a particular substance.)

Considering the public controversy that has attended the cancer threat associated with all these substances at one time or another, the uncertainty toward EDB becomes notable as the exception that proves the rule--virtually all respondents felt capable of judging the cancer threat posed by almost all of these highly controversial substances. Despite this anomaly, those scientists who did rate EDB were most likely to place it at the low end of the scale, as were a majority (52 percent) of those rating DDT and large pluralities of those who rated radon (49 percent) and hormones (45 percent).

The scientists expressed the least concern toward four substances whose cancer-causing potential has generated headlines but which they consensually regarded as only minor contributors to cancer rates. Nuclear power, the pesticide Alar, and artificial sweeteners such as aspartame and saccharin generated mean ratings ranging from only about 1.2 to 2.5. All were regarded as minor cancer agents by large majorities, the size of which ranged from 64 percent (Alar) to 83 percent (aspartame). No more than 7 percent of cancer experts rated any of these substances as a major contributor to cancer rates.

SCIENTIFIC CONTROVERSIES

As our earlier review of the cancer policy debate made clear, it is not only the carcinogenic potential of specific substances that is at issue. Just as hotly disputed are the ways in which the threat of environmental cancer can be identified and its overall magnitude determined. As it turns out, however, the scientific controversies that resound the loudest in the policy arena are considerably more muted within the expert community.

Our survey polled expert opinion on some of the most publicly visible scientific controversies involving the source, detection, and extent of environmental cancer risks. Do we face a cancer epidemic in this country? Are carcinogens unsafe at any dose? Can human cancer risks be inferred from animal tests? And should a single finding of carcinogenicity be sufficient cause for banning a substance from human use? The results appear in figure 4.1.

Is cancer so prevalent in contemporary American society that it can be said to have reached epidemic proportions? The notion of a cancer epidemic has surfaced repeatedly in public discussion of environmental health risks, at least since Joseph Califano's highly publicized 1978 prediction of a surge in occupational cancers. It often appears in conjunction with concerns that the products (and byproducts) of industrial activity, such as pollution and artificial chemicals, are responsible for a long-term rise in cancer rates over the past several decades.

In our survey, however, two out of three cancer researchers (67 percent) rejected the notion of a cancer epidemic in America, and only 31 percent assented to it. In answer to a separate question that asked whether rising cancer rates mainly reflect the results of industrial activity or the combined effects of tobacco and aging, two out of three scientists (65 percent) selected the latter, compared to only 15 percent who blamed industrial activity. Another 15 percent responded that both were equally to blame, and 5 percent were unsure.

Are cancer-causing agents unsafe at any dose? Respondents in our survey of experts disputed that assertion by a margin of more than two to one (64 percent to 28 percent, with the rest unsure).

Can the results of animal studies of suspected carcinogens be extrapolated to humans in order to assess the health risks associated with specific substances? This has been one standard procedure for establishing carcinogenicity in accordance with federal regulations. This approach has long been controversial, however, since it involves giving very high doses of a substance to animals and projecting the results to humans, who are exposed to far lower doses of the same substance. When we put this question to the expert community, the controversy seemed closer to a consensual rejection of these procedures and the assumptions they involve. Only one in four cancer researchers (27 percent) endorsed the practice of assessing human cancer risks by giving animals what is termed the "maximum tolerable dose" of suspected cancer-causing agents. More than double that number (63 percent) disagreed, with the remainder unsure.

Figure 4.1 *Cancer Controversies: Scientists Survey*

Rising cancer rates mainly reflect:

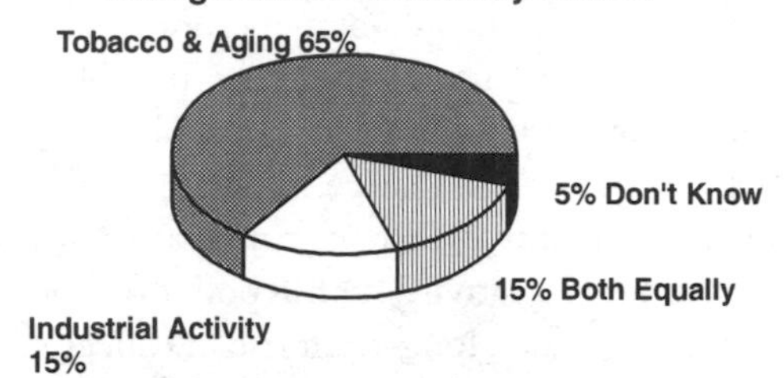

Should chemicals and additives be banned from food and drugs if they ever cause cancer in any species?

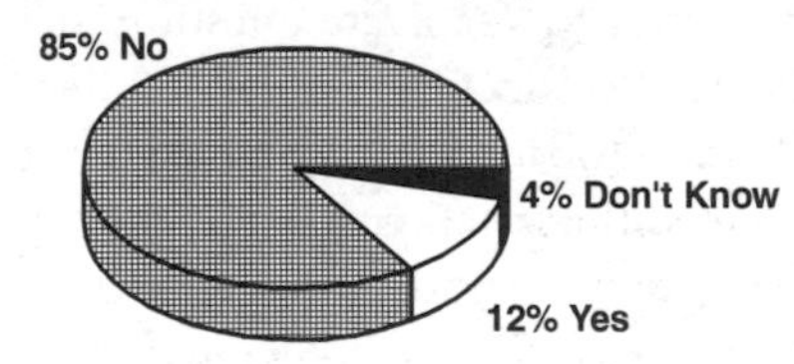

Does the United States face a cancer epidemic?

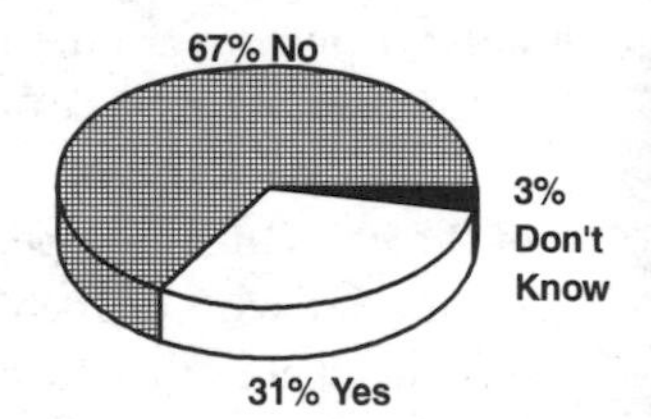

Are cancer-causing agents unsafe regardless of the dose?

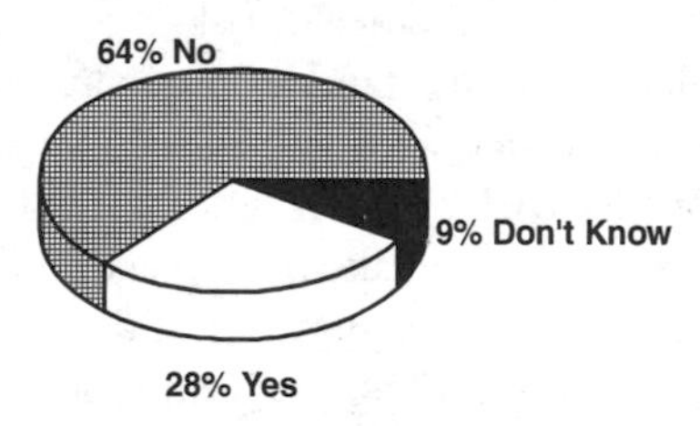

Should human cancer risks be assessed by giving animals the maximum tolerable dose of a suspected cancer-causing agent?

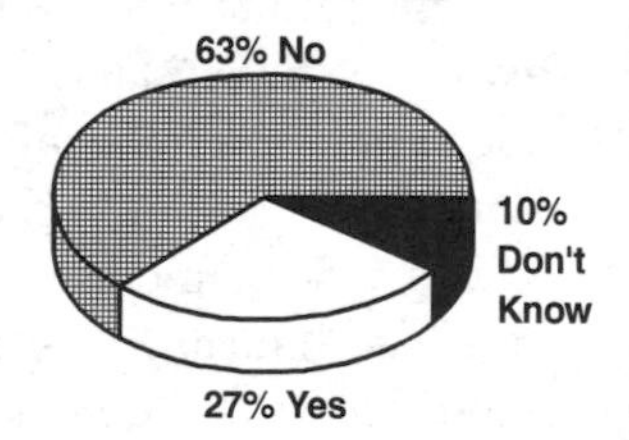

Source: 1993 survey of 401 members of the American Association for Cancer Research.

Finally, we asked scientists to evaluate the zero-risk standard embodied in the now repealed Delaney clause, which held that chemicals and additives must be banned from food if they ever are shown to cause cancer in any species. This principle had sparked public controversy ever since it had led to the FDA's decision to ban saccharin in 1977. But the surveyed researchers rejected this principle by an overwhelming margin of seven to one (85 percent to 12 percent, with the rest unsure).

In summary, large majorities of the research community reject as overly risk-aversive several propositions and practices that currently guide environmental cancer policy. Most cancer experts dismiss the popular notion of a cancer epidemic in America, and they attribute the observable rise in cancer rates to tobacco use and aging rather than the products of modern industry.

Most researchers also reject some of the principles that underlie the current regulatory approach to environmental cancer. These include the inference of human cancer risk from animal tests involving high dosages of suspected carcinogens, the idea that cancer-causing agents are unsafe at any dose, and the analogous zero-risk regulatory standard for evaluating food and drugs.

THE 1984 SMITH COLLEGE SURVEY

Not only are these results internally consistent with regard to the cancer threats posed by environmental agents, but they appear to be stable over time. A similar survey of cancer experts nine years earlier produced findings that parallel those of the 1993 sample. This study, whose results appear for the first time here, was conducted by Stanley Rothman in 1984 and 1985 under the auspices of the Center for the Study of Social and Political Change at Smith College.

For this earlier (and heretofore unpublished) survey, a questionnaire was mailed to scientists who were randomly selected from membership lists of both the AACR and the American Society for Clinical Oncology, which is made up of doctors who treat cancer patients. Thus, the sample represented both cancer researchers and medical specialists who apply this research in treating the disease. Out of 800 researchers and practitioners who were contacted, 457 filled out the questionnaire, a response rate of 57 percent.

Table 4.3
Substances as Possible Causes of Cancer: Comparison of 1993 and 1984 Surveys of Scientists (Ratings on 0–10 Scale)

	Mean Scores		Rank		Percent Rating as Major[a] Cancer Cause	
	1993[b]	1984[c]	1993	1984	1993	1984
Tobacco	9.21	9.27	1	1	96%	98%
Asbestos	6.49	6.88	2	2	56	64
Unhealthy diet	6.38	6.27	3	3	52	59
Sunlight	6.33	6.07	4	4	54	55
Fat in diet	5.39	5.32	5	5	33	42
Aflatoxin	4.85	3.95	6	11	34	24
Sexual behavior (STDs)	4.74	4.40	7	9	28	27
Dioxin	4.74	4.32	8	10	33	29
Pesticides/herbicides	4.72	4.95	9	8	26	37
Pollution	4.70	5.34	10	6	26	44
Alcohol	4.59	5.05	11	7	17	34
EDB	4.22	3.84	12	12	17	23
Infection	3.96	3.16	13	15	14	12
DDT	3.83	3.23	14	14	21	16
Food additives	3.27	3.66	15	13	11	17
Saccharin	1.64	2.12	16	16	3	5

[a] *Major = rating of 7 to 10*

[b] *1993 Study: Random sample of American Association for Cancer Research members specializing in carcinogenesis or epidemiology; N=401*

[c] *1984 Study: Random sample, combined membership of AACR and American society for Clinical Oncology; N=457*

Most important for our purposes, this earlier survey used the same 0-to-10 scale to rate many of the substances included in the Roper survey. This permits us to treat our survey findings as a direct replication of results obtained nearly a decade earlier, a rarity in elite survey research. A comparison of the ratings given by the two expert samples appears in table 4.3.

The results are notable for the similarities of both the average scores and the relative rankings assigned to the various cancer agents. The two groups' top five rankings are identical, and the mean scores assigned to each substance are nearly so. Out of sixteen substances/conditions whose contribution to cancer rates were rated by both groups, only five differed by as much as one-half point on the ten-point scale, and none differed by a full point.

Tobacco, asbestos, diet, and sunlight top the list of carcinogens for both samples. Conversely, EDB, DDT, food additives in general, and saccharine appear at or near the bottom of both lists. The expert community currently assigns slightly more significance to the dangers of aflatoxin, DDT, and infection, while their counterparts of a decade ago were somewhat more concerned about pollution. The ratings for dioxin, pesticides, and food additives fell below the midpoint on the scale for both groups, while pollution barely surpassed that mean rating for the earlier sample only.

The earlier survey also addressed some of the same scientific controversies that are discussed above, although the wording of questions varied somewhat in the two surveys. Nonetheless, the responses uniformly supported our finding of scientific skepticism toward some widely held assumptions in the cancer policy debate. For example, in the 1984 survey, 71 percent disagreed that the United States is "experiencing an epidemic of cancer," virtually identical to the more recent result.

Even larger majorities rejected the zero-risk standards and assumptions that are implicit in much federal cancer policy and explicit in the Delaney clause. Thus, an overwhelming 92 percent rejected the notion that there is no safe threshold of risk for carcinogens. A nearly identical 91 percent disagreed with the Delaney clause principle that food and drug products should be banned on the basis of one positive result of carcinogenicity from animal tests. Given these findings, it is rather surprising, to say the least, that the clause remained relatively unchallenged for so long.

Considering the differences in sampling procedures and questionnaire administration, the similarities in these two sets of survey findings are striking. They strongly reinforce our argument that the broad contours of scientific opinion in this area have changed relatively little in recent years. Most researchers have long ranked artificial chemicals and industrial by-products relatively low among environmental cancer risks, just as most have disputed some popular assumptions about the dangers posed by cancer-causing agents in the environment.

Among the highly publicized cancer scares that have attended such widely feared substances as DDT, EDB, dioxin, and other pesticides and additives, only asbestos appears to have elicited commensurate concern from the experts. According to our expert sample, the other substances have received attention disproportionate to their standing in the hierarchy of health risks established by medical science.

Thus, the views of cancer experts have been consistent and relatively unchanged over the past decade. Moreover, these portraits drawn from systematic surveys accord well with the more traditional literature reviews presented in chapter 1. We may infer that no startling new evidence has arisen in the past two decades to alter significantly experts' collective opinions on cancer risks and causation. It seems reasonable to assume that the disjunction between popular perceptions and scientific understanding of environmental cancer is one of long standing.

EXPERTISE AND IDEOLOGY

A different kind of objection may be raised against this approach to contrasting popular with expert opinion. In such a contentious and politicized debate, do the experts' perspectives necessarily reflect their superior knowledge? Or might even expert opinion be distorted by self-interest or partisan commitment?

Could there be, for example, an ideological or partisan tilt among these scientists that influences their opinions? Or is the sample of cancer scientists perhaps weighted toward private-industry researchers who have direct or indirect pecuniary interests in the man-made substances under consideration?

Any such concerns should be alleviated by the background information on cancer scientists presented in figure 4.2. Insofar as the political debate over cancer policy breaks along partisan lines, the most

Figure 4.2 *Backgrounds of Cancer Scientists*

Ideology

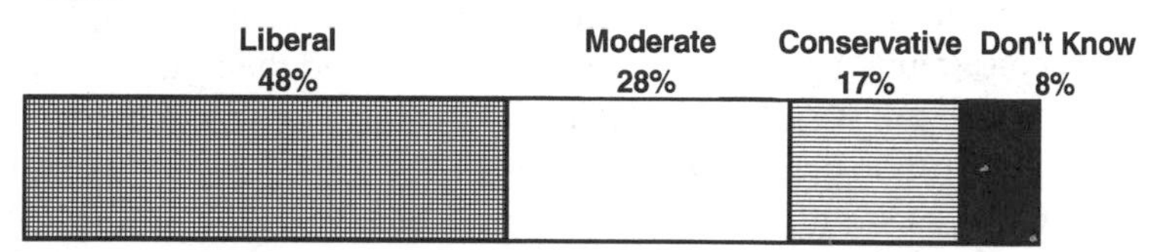

Political Party

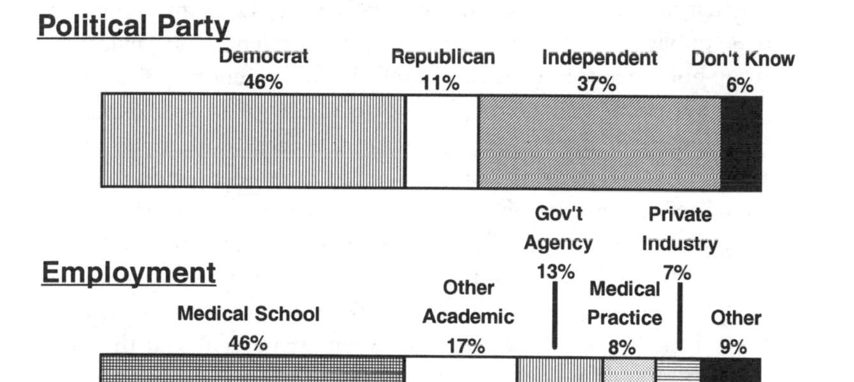

Professional Activity

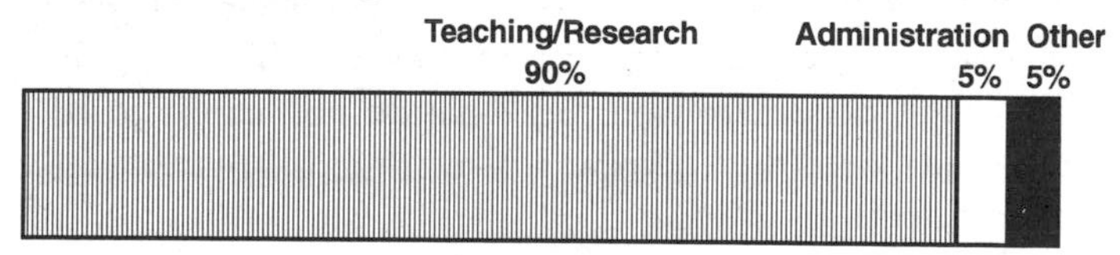

Source: 1993 survey of 401 members of the American Association for Cancer Research.

risk-averse and regulatory-friendly perspectives have come from the ranks of liberal Democrats and their allies in the labor and environmental movements, while these positions have been attacked most consistently by conservative Republicans and their business allies.

When we asked cancer researchers about their own ideological and partisan orientations, however, they turned out to fall mainly in the liberal and Democratic camps. Indeed, self-described liberals outnumbered conservatives among this group by nearly a three-to-one margin, and there were more than four times as many Democrats as Republicans. This profile makes cancer scientists a left-leaning group both in absolute terms and relative to the general population.

The liberal and Democratic orientation of cancer experts is consistent with their professional status as faculty members at research universities. Nine out of ten listed their occupation as either researcher or academic faculty, and nearly two out of three (63 percent) were based at a medical school or other academic institution. By contrast, only one in fourteen was employed by private industry. In fact, nearly twice as many worked in government agencies as worked in the private sector. Moreover, 73 percent said they had never consulted for private industry, and only 8 percent had done so more than three times.

In addition, there was no consistent differentiation along political or ideological lines in the views these researchers expressed on the cancer risks and scientific controversies described above. As a group, the experts—whether conservative or liberal, Democratic or Republican—viewed cancer risks along roughly the same lines. Thus, their perspectives on this topic do not appear to be "contaminated" by either narrow self-interest or broader ideological commitments.

This is hardly an unexpected finding for a survey of scientific researchers who are asked about their area of professional expertise. But it contrasts with the structures of belief among lay publics, whose opinions on scientific matters frequently reflect broader social dispositions or demographic factors. Moreover, as on other environmental controversies, public debate over environmental cancer has often taken on overtones of economic interest and political partisanship.

Journalists and legislators frequently sort out competing claims in this policy arena by turning to researchers and other representatives of a few well-known nonprofit organizations that have emerged as leading advocates of environmental protection. (We discuss this relationship at length in chapter 5.) To probe more deeply into the relationship between scientific opinion and public policy, therefore, we turned from the experts to the activists—the leaders of the environmental movement.

Environmental Activists on Cancer Causation

Environmental groups are relative newcomers to the health policy debate. The conservation movement traces its roots back to the nineteenth century. But it took the social turbulence of the 1960s to forge an alliance of traditional nature and wildlife preservation groups with newer, more activist groups whose concerns stretched from the energy sources buried deep within the earth to the ozone layer high above it.

Thus, the environmental movement has come a long way in a short time. A quarter-century since it emerged as a self-conscious, grass-roots protest movement, environmentalism is now thoroughly in the mainstream. A social movement that challenged entrenched institutions has become part of most public school curricula, with its own national celebration (Earth Day) and even its own cartoon series (Ted Turner's "Captain Planet"). Other sure-fire indicators of mainstream status include cries of "sellout" from longtime activists, support from corporations anxious to join the bandwagon, and a backlash in the form of the so-called property rights movement.

For most ordinary Americans, environmentalism has become what political scientists call a valence issue—a mom-and-apple-pie value to which just about everyone gives lip service. In their responses to pollsters, most Americans place environmental concerns above even economic security, and large majorities proclaim support for "whatever it takes" to protect the environment (Ladd and Bowman 1995). On a rhetorical level, at least, it is possible to proclaim, with only mild hyperbole, that we are all environmentalists now.

Environmentalists have had remarkable success in winning both the high ground and the grass roots away from their wealthy, well-established, politically connected opponents in corporate America.

This political success rests largely on their ability to embody a combination of intellectual and moral authority. Environmentalists come across as people who both know about the environment and care about it, while their opponents seek to use (or abuse) the environment to further their own financial interests.

It is in this sense that environmental activist groups have emerged as a major wing of the "public interest" movement. They profess to advocate our common interest in preserving the planet. This mission differentiates them from the vested or special interests, whose aims are held to be less benevolent or disinterested. It is also the major source of their public influence, to the (considerable) degree that journalists present their opinions as dispassionate assessments that transcend self-interest and partisanship.

In an area as complex as environmental cancer, however, purity of motive is not sufficient to produce sound policy. Accurate information and scientific expertise are prerequisites for any rational response to environmental health hazards. This is why the major environmental groups employ staff scientists, conduct studies, and issue technical reports in conjunction with their policy prescriptions.

But anyone's understanding of complex social phenomena can be influenced by underlying perspectives and ideological commitments, as well as economic interests. Environmental activists are no exception to this rule, as we found from a survey of the movement's leadership. This survey, which was conducted independently by the Center for Media and Public Affairs, included questions about environmental cancer as part of a broader inquiry into the social and political beliefs of environmental leaders.

POLLING THE ACTIVISTS

To poll the leaders of the environmental movement, we first identified the most prominent organizations, on the basis of major-media mentions. A dozen groups have received the most coverage in recent years for their environmental or conservation activities: the Clean Water Action Project, Ducks Unlimited, the Environmental Defense Fund, Friends of the Earth, Greenpeace USA, the National Audubon Society, the National Wildlife Federation, the Natural

Resources Defense Council, the Nature Conservancy, Renew America, the Sierra Club, and the Ralph Nader-affiliated U.S. Public Interest Research Group.

On the basis of annual reports and other published information, we identified 442 individuals who serve as these groups' senior staff members, heads of state or regional chapters, and on national boards of directors. One hundred of these environmental leaders, selected randomly from the list of 442, were interviewed by telephone from July through September 1993. The interviews were conducted by the Roper Center for Public Opinion Research. The Roper Center also assisted in designing the survey to permit direct comparisons with the general public by asking questions identical to those used in recent national polls.

Among a wide range of questions about environmental matters, these activist leaders were asked to rate the cancer risk associated with many of the same substances about which cancer researchers were also queried. The exact same question wording was used to permit direct comparisons between the ratings of activists and scientists. The results for both groups are shown in table 4.4.

Environmental leaders assigned higher risks than cancer researchers to eleven out of thirteen substances listed. Only tobacco and sunlight attracted slightly more concern from the researchers. In the case of man-made substances, the differences in their ratings were frequently dramatic. At least twice as many activists as scientists detected "major" cancer threats from Alar, artificial sweeteners, DDT, dioxin, food additives, and nuclear power.

The differences were even greater at the other end of the scale, with scientists far more likely than activists to rate most substances as "minor" causes of cancer. Thus, about twice as many researchers as environmentalists rated Alar and nuclear power as minor carcinogens; three times as many researchers placed DDT and asbestos at the low end of the scale; five times as many researchers regarded food additives and dioxin as minor threats; and a whopping seventeen times as many cancer specialists as environmental leaders (by 34 percent to 2 percent) saw pollution as a minor contributor to cancer rates.

Another difference between the activists and the scientists was

Table 4.4
Comparisons of Substances as Causes of Cancer by Environmental Activists and Cancer Scientists (Ratings on 0–10 Scale)

	Activists[a]		Scientists[b]	
	Mean	% Major[c]	Mean	% Major
Smoking	9.1	85	9.2	95
Dioxin	8.1	67	4.7	33
Asbestos	7.8	63	6.5	56
EDB	7.3	28	4.2	17
DDT	6.7	47	3.8	21
Pollution	6.6	40	4.7	26
Sunlight	6.3	41	6.3	54
Fat in Diet	6.0	39	5.4	33
Food additives	5.3	19	3.3	10
Nuclear plants	4.6	22	2.5	7
Alar	4.1	16	2.2	6
Saccharin	3.7	12	1.6	3
Other sweeteners	2.8	6	1.2	1

[a] *Activists: Survey of leaders from 16 nationally prominent environmental and conservation groups, taken July–September 1993, N=101*

[b] *Scientists: Random sample of American Association for Cancer Research members specializing in carcinogenesis or epidemiology, taken January–February 1993; N=401*

[c] *Major = rating of 7 to 10*

the level of perceived knowledge and certitude they brought to their respective assessments of cancer risks. The scientists, on all but a few matters, were confident of their expertise; 0–7 percent responded "don't know" in twenty-seven of the thirty cases in tables 4.1 and 4.2. In only one case (EDB) were more than 20 percent of scientists surveyed unwilling to venture an informed estimate. The activists in every case registered at least 10 percent in the "don't know" category, with more than 20 percent lacking sufficient knowledge on eight of the twelve questions.

When asked about some of the major controversies in cancer incidence and causation, the environmentalists again disagreed with the scientists by large margins. These comparisons appear in figure 4.3. Only one-quarter (27 percent) of activists rejected the notion that the United States faces a cancer epidemic, compared to two-thirds (67 percent) of cancer scientists. More than twice as many activists as scientists also held industrial activity responsible for rising cancer rates (activists, 64 percent; scientists, 30 percent).

If environmental leaders were more liberal than scientists in their estimates of the cancer threat, they were more conservative in the standards they applied to dealing with cancer risks. Three times as many activists supported the Delaney clause standard that bans the intentional addition to food of any substance ever shown to cause tumors in any species (activists, 39 percent; scientists, 12 percent). Just less than half of activists (49 percent) disputed the zero-tolerance principle that holds cancer-causing substances to be unsafe regardless of the dose, which 64 percent of scientists rejected.

To compare researchers and activists with more precision, we developed a risk perception scale for man-made substances that are generally perceived as cancer-causing agents about which controversy swirls today, such as Alar, nuclear plants, EDB, and pesticides. (Tobacco and asbestos were not on the scale.) Possible scores ranged from 1 to 10. On the scale, scientists scored 3.40; environmental activists scored 5.61, a difference of more than two points. A similar scale of "natural" carcinogens (such as low-fiber and high-fat diets and sexually transmitted disease) revealed a much smaller difference. Clearly environmental activists are not simply risk-aversive.

A number of other findings are of interest. Self-identified ideology plays only a marginal role in explaining differences in scores among researchers. Interestingly, the more conservative the researcher self-identifies (on a conservative-to-liberal scale), the higher he or she rates the risk. The few scientists employed by industry do score marginally lower than university- or hospital-employed researchers, as do those few who have consulted with private industry as compared to those who have not. Female scientists score more than a point higher than male scientists in estimating risk. On the other hand, the more expert a scientist as measured by the number of peer-reviewed articles he or she has published, the lower his or her risk score. Those scientists who had published a substantial number

Figure 4.3 *Environmental Activists vs. Scientific Experts on Selected Environmental Issues*

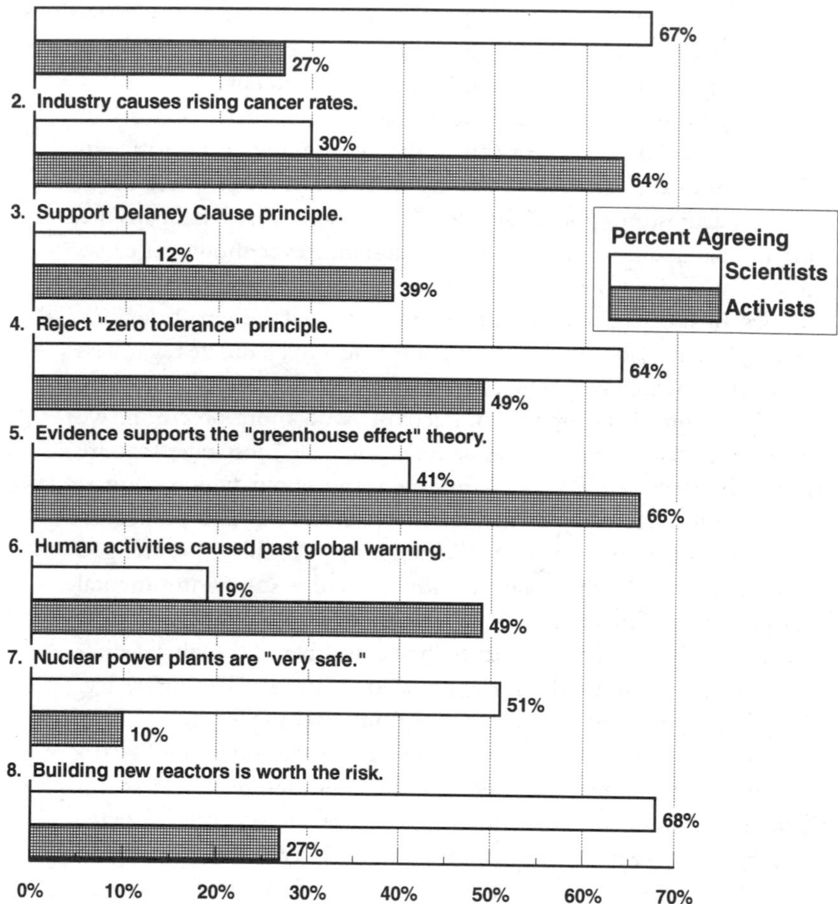

Scientific Samples: 1–4. Roper survey of 401 cancer scientists, 1993; 5–6. Gallup survey of 400 climate, atmospheric, and oceanographic scientists, 1992; 7. Center for Media and Public Affairs survey of 358 energy scientists, 1985; 8. Yankelovich survey of 418 "leading scientists," 1992.

of articles scored a full point lower than those who published no articles, and this figure probably underestimates the difference.[4]

The environmental activists are a much more homogeneous group. We found no significant statistical differences among them, though again women activists seem to be more risk-aversive than male activists, scoring one point higher.[5]

The thread that runs through these differences is a consistent tendency by environmentalists to magnify the risks of environmental cancer beyond the estimates of scientific researchers in this field. Apparently, this reflects a more general tendency by this group to perceive environmental threats of all sorts as greater than the relevant academic experts do, especially threats from man-made cancer-causing agents as against those produced by nature. This conclusion derives from their responses to several other survey questions, each of which repeated the exact same phrasing that was used in previous surveys of technical experts.

Thus, in addition to rating cancer risks as more dire than cancer researchers did, the environmental leaders were more certain about the threat posed by global warming than were researchers in that field. In this case the comparison comes from a 1992 Gallup poll of four hundred climatologists, oceanographers, and atmospheric scientists who were randomly selected from their professional associations. Among these experts in climate change, only two out of five (41 percent) believed that current scientific evidence supports the existence of a human-induced greenhouse effect. By contrast, two out of three environmentalists (66 percent) affirmed this position. Similarly, only one in five climate researchers (19 percent) attributed past increases in global temperature to human activity, compared to half (49 percent) of the environmentalists.

As noted in chapter 2, a scientific consensus that the earth is warming is growing stronger, though substantial disagreement still exists as to what role is played by human activities. Nevertheless, as indicated by a United Nations–sponsored panel in 1995, a larger proportion of experts in the field is probably convinced at this point that human activities do play a role than agreed with that proposition when the survey was conducted some years earlier. Also, estimates as to the degree and rapidity of climatic change suggested by the UN panel report are lower than had been claimed by the more ardent supporters of the warming hypothesis (See Stevens 1995a, 1995b, 1996).[6]

In light of these findings, it should occasion no surprise to learn that these same environmental leaders also viewed nuclear energy as more dangerous than did energy scientists. Here the evidence comes from several surveys. In one study 358 energy scientists, randomly selected from *American Men and Women of Science,* were asked to rate the safety of nuclear power plants. They were first surveyed in 1980, soon after the accident at Three Mile Island, and were resurveyed in 1985, after the Chernobyl disaster. In both years a majority rated plants as "very safe" and more than 80 percent rated them as at least "moderately safe." By contrast, only one in ten environmental leaders rated nuclear plants as very safe, and a minority (44 percent) thought they were at least moderately safe.

In sum, these leading environmental activists rated a variety of environmental risks as greater than scientific experts did. This reinforces the conclusion that whatever the merits of their arguments, environmental activists do not represent a source of expert opinion on issues of environmental cancer risks and causation.

If the green lobby does not speak for the cancer experts, for whom do these activists speak? It might be argued that they represent public concern over environmental threats, which should be taken seriously regardless of expert opinion. As we noted earlier, numerous polls show strong public support for environmental protection policies that reflect a mentality of "better safe than sorry."

As with many such valence issues, however, the public's united front with the issue activists begins to slip when specific cases and hard choices are presented (*American Enterprise,* 1995). Figure 4.4 shows some comparisons in which environmental leaders responded to the same questions that were used in recent national surveys. They show widespread differences between the activists and the public on some contentious issues of public policy.

Thus, although various polls find public support for more spending on environmental protection, only one-third of the public would increase taxes to that end. (Data on public attitudes toward environmental issues come from a 1992 Roper poll.) The environmentalists are almost twice as likely as the general public (by

Figure 4.4 *Environmental Activists vs. General Public*

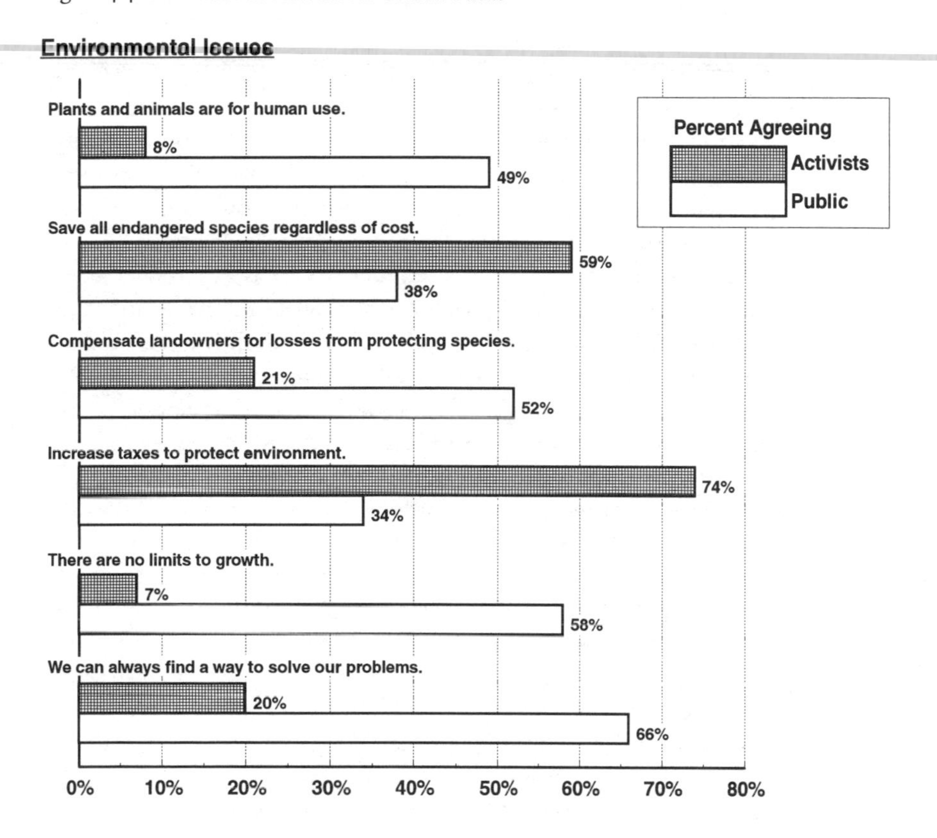

Activists: Survey of 101 leaders from nationally prominent environmental and conservation groups, taken July – September 1993.

Public: Taken from 1993 national surveys by Gallup, Roper, and CBS/*New York Times*. Question wording was identical to the activist survey.

74 percent to 34 percent) to back up their professed concerns with their pocketbooks.

Similarly, three out of five environmental leaders (59 percent) believe that all endangered species should be saved regardless of cost, a commitment that is shared by only 38 percent of the public. Conversely, a majority (52 percent) of the public would compensate landowners for economic losses incurred in protecting wetlands, while only one in five activists (21 percent) would do so.

The differences go far beyond pocketbook issues. Indeed, they extend to some core philosophical beliefs about the relationship between the human environment and the physical environment. Most environmental leaders reject the traditional American optimism that problems can always be solved, growth can continue unabated, and the environment should be viewed primarily as a resource to assist human endeavors. But the bulk of the public still holds to this anthropocentric outlook, though their responses do vary with the kind of questions asked.

Continuing to contrast our own survey with national Roper data, we found that the public was a stunning eight times more likely than the activists—by 58 percent to only 7 percent—to believe that there are no limits to growth in America today. And by a six-to-one margin (49 percent to 8 percent) the public was more likely to regard plants and animals as existing "primarily for human use."

It is not that environmental activists perceive greater dangers or harbor a greater sense of urgency than the public at large. The activists and the national sample were both about evenly divided over the notion that "the 1990s is the last decade when humans will have a chance to save the earth from environmental catastrophe." Fifty-two percent of the activists agreed, and so did 49 percent of the general public. Rather, it appears that the average American is simply more sanguine about our ability to avoid these perils. By a margin of 58 percent to 43 percent the public outpaced the activists in asserting that "technology will find a way of solving environmental problems." More generally, two out of three citizens overall—but only one in five environmentalists—believe that "we can always find a way to solve our problems and get what we want."

Thus, the activists' perspectives on environmental issues reflect neither any special scientific or technical expertise nor the concerns of the public at large. Both these conclusions are reinforced by the social backgrounds of the activist sample. Few possess advanced scientific training. Only one out of four holds a college degree in science or engineering. On the other hand, one-half are lawyers or social scientists by training.

Nor are environmental activists representative of the general public. Three out of four (76 percent) are male. An overwhelming 97 percent of those responding are white, 2 percent are Asian, and only 1 percent are black. They are also far more affluent than the average American. More than one out of four (27 percent) of those responding have annual family incomes above $100,000, and the vast majority (78 percent) earned more than $50,000.

What this demographic portrait fails to reveal is the activists' distinctive socio-political perspective. Their environmental commitments are congruent with a cluster of liberal attitudes ranging from progressive racial and sexual attitudes to antimilitarism to faith in government regulation. Once again, their distance from the general public on these dimensions can often be expressed in multiples. As shown in figure 4.5, environmental leaders were more than three times as likely as the public to identify themselves as political liberals (63 percent of activists versus 18 percent of the public) and only one-fifth as likely to be Republicans (6 percent versus 31 percent). There were ten times as many Democrats as Republicans (59 percent versus 6 percent) among environmental leaders, at a time when the public was about evenly divided between the two parties (35 percent Democrats versus 31 percent Republicans). (Note that these data come from 1993 surveys, taken well before the surprise Republican political ascendance in the 1994 elections.)

These political differences were matched by those of religious orientation. Nearly half the activists (47 percent) were nonreligious, compared to only 6 percent of the general population. Similarly, only one out of eight activists (12 percent) described themselves as regular churchgoers, compared to 42 percent of the populace. (Figures for the public come from 1992 surveys; data on religious affiliation and observance have been relatively stable in recent years.)

Perhaps most telling are the public and activist attitudes on social issues. Environmental leaders are twice as likely as the average American to support abortion rights for "parents [who] don't want

Figure 4.5 *Environmental Activists vs. General Public*

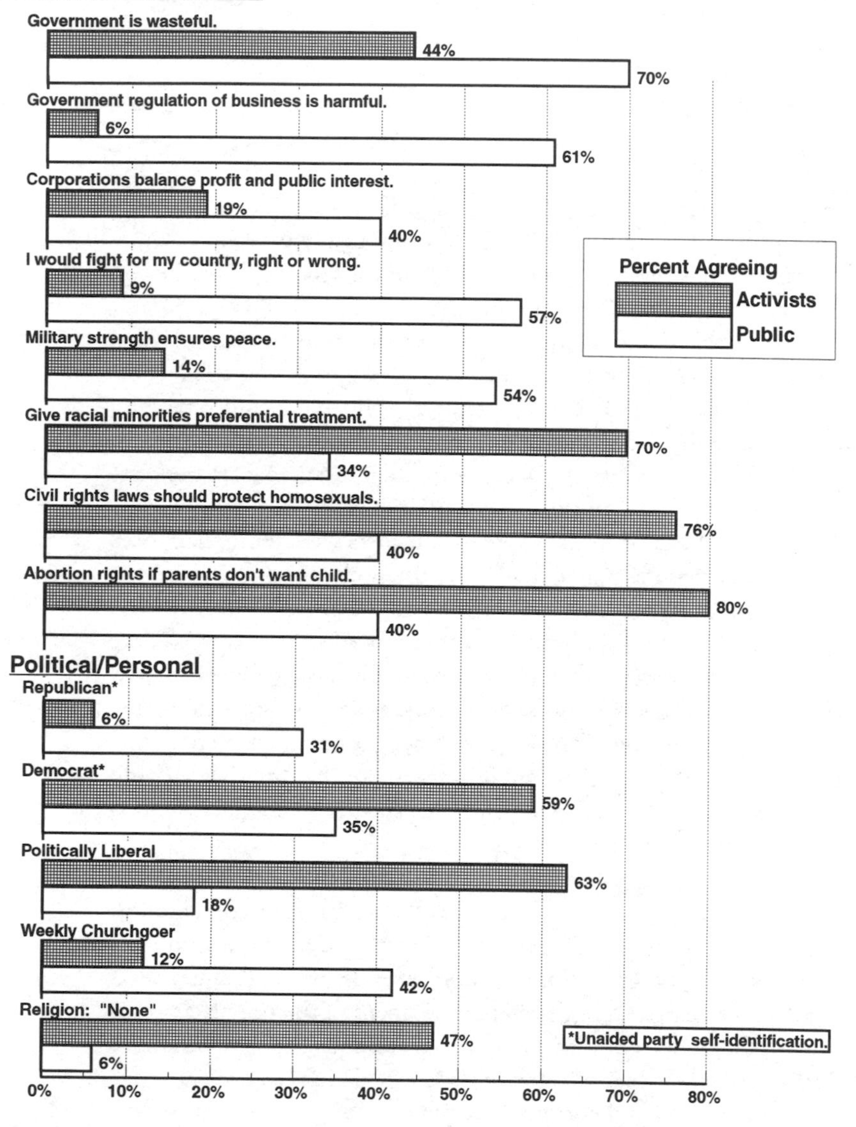

another child" (80 percent versus 40 percent), "preferential treatment" of racial minorities (70 percent versus 34 percent), and the use of civil rights laws to protect homosexuals (76 percent versus 40 percent).

The public is four times as likely as the activists to believe that "military strength ensures peace" (54 percent versus 14 percent), and six times as many members of the national sample would be willing to "fight for my country, right or wrong" (57 percent versus 9 percent). The environmentalists' antimilitarism is nearly matched by their suspicion of business. Whereas 40 percent of the public agree that corporations "balance profit and the public interest," fewer than half as many activists (19 percent) concur.

Conversely, the environmental leaders were far more willing than the public at large to put their trust in government. At a time when 70 percent of the public agreed that "government is wasteful," only a minority (44 percent) of activists endorsed that widespread criticism. Most striking of all, the public was more than ten times as likely as the environmentalists to criticize government regulation of business. Sixty-one percent of the populace were willing to characterize government regulation as "harmful," compared to only 6 percent of the activists.

From these data it seems clear that environmental activists do not speak for the American public. Although most Americans are willing to describe themselves as "environmentalists," the perspectives and backgrounds of this movement's leadership are considerably removed from those of the majority. If the activists don't represent the public consciousness and don't speak for (or from) expert opinion on scientific matters, precisely where do their views on environmental health threats come from?

It seems reasonable to conclude that the views of environmental activists proceed from a broad ideological position. They have assumed for themselves the mantle of advocates for the public interest with regard to environmental issues. But their characterization of the public interest, and the public policies that it requires, proceeds from a familiar strain of social and cultural liberalism that is often identified with upper-middle-class professionals.

None of this is intended to suggest that environmental leaders are insincere in their beliefs or that their policy prescriptions should be accorded less attention or respect than those of any other interest group seeking to influence the national agenda. One may recognize the social and political distinctiveness of this group without endorsing the caricature of tree-hugging urbanites shod in Birkenstocks. As *Newsweek* editor Gregg Easterbrook, who considers himself a friendly critic of the environmental movement, has noted, "Politically, environmentalism has become a refuge of anti-growth and counterculture sentiment . . . a sense that material culture has lost its bearings, and that awareness of our roles in nature may cure what ails modern humanity" (Easterbrook, 1990, p. 18).[7]

But this profile does help to explain the movement's success in getting its message out in a media environment clogged with competing claims about cancer risks. The media's receptivity to the messages of environmental groups is based less on their representativeness or their expert knowledge than on the moral suasion exerted by their vision of the public good. It is a vision that seeks to overcome the brutality of the marketplace and the indifference of bureaucracy through the goodwill and selfless efforts of individuals who put the planet's interests before their own.

Theirs is a vision that has much in common with another group that regards itself as the public's trustee, standing above the battle of vested interests to protect the commonweal. Journalists see themselves as partisans for the public at large against the powers that be, in a manner similar to the environmentalists' chosen role of partisan for the planet. And in this age of media politics, the advantage would seem to go to those who share the sensibilities and cultural affinities of major media journalists, judging from the evidence we present in chapters 5 and 6.

5
Media Coverage of Environmental Cancer

"STUDY LINKS HOT DOGS, CANCER: Ingestion by Children Boosts Leukemia Risk, Report Says" (*Washington Post* 1994a). Families heading off to baseball games and barbecues on June 3, 1994, may have had second thoughts after seeing this story in their morning newspaper. If the headline kindled parental fears, the lead sentence fanned the flames: "Children who eat more than 12 hot dogs per month have nine times the normal risk of developing childhood leukemia, a University of Southern California epidemiologist reports in a cancer research journal." The widely reprinted *Los Angeles Times* wire service dispatch went on to describe elevated risk levels not only for children whose mothers ate hot dogs during pregnancy but also for those whose fathers ate hot dogs *before conception*.

The most remarkable thing about this news report is how unremarkable it was. This was no *60 Minutes* exposé of Alar poisoning. There was no panic over cancer agents seeping through the ground, as at Love Canal, or being dispersed by the wind, as at Three Mile Island. The entire episode barely registered as a tremor on the media Geiger counter. It was just the latest episode in the long-running series popularly known as the cancer scare of the week.

In fact, the article went on to damp some of the fires sparked by the eye-catching lead and headline. Far from being sensationalistic, this news story was notable for including numerous qualifications and dissenting voices. It cited critics who argued that the findings were suspect because people have difficulty recalling what they ate, and that the statistical association was relatively weak and might be caused by other variables. The researchers themselves did not argue that people should stop eating hot dogs, recommending only "a small modification of your consumption." In its final paragraph, the story noted that the results had not been published in a peer-reviewed journal, a sine qua non of scientific acceptability.

Indeed, the conclusions were so heavily qualified that one might ask why this was "news" at all. This question also occurred to someone at the *Washington Post,* which ran a light-hearted follow-up article headlined "Hold the Relish: A Wurst-Case Scenario" (*Washington Post* 1994b). This "Style" section piece sympathized with all the rueful readers who "open the paper and read the headline and . . . [get] your daily dose of fear." It quoted a puzzled National Cancer Institute epidemiologist who had spent much of her day answering media inquiries. After refusing requests for live television interviews, she complained, "I think it's blowing this out of proportion. I don't know why the press got into this."

Why indeed? For all its flippancy, the *Post* essay was the journalistic equivalent of a cry for help: "Stop us before we scare again!" But should the media really be blamed for the spread of "cancerphobia," the fear that everything causes cancer? Or is this just another case of blaming the messenger for bringing bad news, in this instance the unwelcome news of our own mortality? After all, few would dispute that overall, cancer rates are rising, or that many substances in our environment are suspected or proven carcinogens.

Moreover, many journalists would argue that they are fulfilling their role as society's watchdogs by calling attention to potential dangers. They gain audiences and win awards for ringing the alarm or blowing the whistle, not for singing lullabies. So the question needs to be reframed: Is media coverage of environmental cancer an accurate reflection of real-world concerns? Are its emphases proportionate to the dangers involved? How well does it convey scientific knowledge of the causes of cancer and the effects of carcinogens?

These are the questions we shall address in this chapter, by means of a systematic content analysis that compares media coverage of environmental cancer with the body of scientific opinion described in chapter 2.

Studying Cancer News

Our goal was to discover how the national media covered news about the environmental causes (or suspected causes) of cancer ever since the federal government launched its "war on cancer" in 1972. The ensuing years have seen a wide range of newsworthy controversies over environmental cancer, involving such diverse substances as pesticides, food additives, asbestos, hormones, and tobacco; news events such as Chernobyl, Three Mile Island, Love Canal, and Times Beach; the regulatory activities of numerous government agencies, including EPA, FDA, CPSC, and OSHA; and a steady stream of scientific studies from government, industry, and academic institutions.

To reduce this vast body of material to manageable proportions, we selected a representative sample of the most visible reports from television, news magazines, and leading newspapers. Specifically, we examined all news stories on this topic that appeared on the ABC, CBS, and NBC evening newscasts or in *Time, Newsweek,* and *U.S. News and World Report,* as well as stories on the front page of any section of the *New York Times,* the *Washington Post,* and the *Wall Street Journal.* This produced a total sample of 1,206 news items.

We began with the number-one source of news for Americans throughout this entire period—the evening newscasts on ABC, CBS, and NBC. Americans have named television as their leading source of news in response to polls taken as far back as 1963. Until recent years the commercial television networks had virtually no competition as agenda setters for the mass audience. Despite inroads from new media ranging from cable to talk radio, the flagship newscasts of the big-three broadcast networks still reach an unparalleled audience night after night. Although their share of the audience has declined sharply during the past decade, they still reach an average of eight to nine million households apiece, or about 25 million people collectively.

By contrast, notwithstanding huge audience spikes during crises such as the Gulf War, the Cable News Network (CNN) normally reaches only a fraction of that number—perhaps 250,000 households for its *Prime News* program and 500,000 for *Headline News*. Competitors such as *The NewsHour with Jim Lehrer* on PBS and ABC's *Nightline* also reach relatively small (if upscale) audiences. Viewership is similarly reduced, as is news content, during the broadcast networks' morning news shows. Among the networks' prime-time news magazine series, only CBS's *60 Minutes* appeared throughout much of the period that we studied. With a few notable exceptions, these shows only rarely deal with cancer causation. All these news programs can be influential, but none can match the regular and cumulative impact of the network nightly newscasts.

Among print media, recent technological changes have not altered the profession's long-standing pecking order. Studies have confirmed what news watchers know intuitively—that the most influential daily outlets are the *New York Times,* the *Washington Post,* and the *Wall Street Journal,* and the key weekly sources of news summary and analysis are *Time, Newsweek,* and *U.S. News and World Report.*

As in the case of television, new competitors for press influence have arisen in recent years. For example, *USA Today* has become the country's first genuinely national daily paper, and the *Los Angeles Times* has developed a national reputation as the flagship of the Times Mirror chain. Whether unfairly or not, however, the former is still struggling for journalistic respectability after being derided as "MacPaper" for its short and snappy, fast-food approach to the news, and the latter has attained *primus inter pares* status among regional papers without quite breaking into daily journalism's big three.

Given the impossibility of examining all media coverage over more than two decades, our sampling approach has several advantages. First, it represents the national media outlets that were most influential across the entire time period, 1972–1992. Since these industry leaders also set the standards for other journalists, their coverage serves as a proxy for the news agenda in many local newspapers and television newscasts. Further, our approach insures that all highly prominent stories are represented. We examined the stories that were most likely to affect the widest audience. A random sample would not differentiate between front-page news and stories that were buried on the inside pages.

Although limiting the newspaper sample to front-of-section stories precludes generalizations about their depth of coverage, it increases the comparability of various types of media. The limited time available in a television newscast forces electronic reporters to be more selective in story choice. Similarly, weekly magazines must limit their story selection to the most high-profile events and issues on the news agenda. To compare these rather finicky formats with the gluttonous news appetite of the daily papers would lead to lop-sided and erroneous conclusions about coverage of cancer issues. Thus, limiting the newspaper sample to the most prominently featured stories more closely approximates the coverage that the news magazines and television networks can provide.

To examine this material, we employed the social-scientific technique of content analysis, which produces systematic and objective descriptions of communicative material. A systematic approach requires that media content and analytic categories be included or excluded according to consistently applied rules. Objectivity involves following explicit rules and procedures that define analytic categories and criteria to minimize a researcher's subjective predispositions. It implies reliability: that is, other researchers applying the same procedures to the data should reach the same conclusions.

Our study proceeded in two phases. First, a qualitative or emergent analysis identified the topics, themes, symbols, stylistic devices, and so on, that made up the form and substance of the news stories. This approach allowed our researchers to develop analytic categories based on examination of the material rather than imposing *a priori* categories on the news. This process insured the best fit between categories and the material being coded. Moreover, this qualitative analysis sensitized us to differences in presentation styles among the news outlets. This project was aided by an earlier examination of the stories under the direction of Stanley Rothman and William Lunch, which identified many key facets of coverage that were later incorporated into this study.

Once the analytic categories were developed and tested, they formed the basis of the quantitative analysis. This involved the systematic coding of story content into discrete categories that provide for valid, reliable, and quantifiable judgments. Reliability checks and measurements were an integral part of this process. Two coders independently applied the system to the same news stories. Only those variables on which they agreed more than 80 percent of the time, a widely accepted threshold, were retained in the final analysis.

Because this process creates systematic and reliable data, it permits researchers to transcend the realm of impressionistic generalizations, which are prone to individual preferences and prejudices. Throughout the coding process, coders were randomly assigned stories to further reduce systematic error. These procedures insure the most objective classification of media content that the social sciences have devised.

THE CANCER STORY

Once the system of content analysis was developed and tested, we applied it to media coverage of environmental cancer. Of course, the cancer story is really several stories covering a shifting base of scientific research and policymaking. The tone and content of coverage at any single point in time often bore only limited resemblance to coverage a few years down the road. Therefore, before discussing the overall pattern of findings, it is useful to glance at a road map of the media's progress along the byways of cancer news.

In the early years media coverage focused on new understandings of viruses and cancer, as well as the risks of manmade chemicals. In the late 1970s attention was fixed on man-made chemicals and pesticides as causes, with other issues dropping into the shadows. In the early 1980s new genetic discoveries vied with concerns over chemical pollution for news space. By the late 1980s attention was turning to diet and aspects of life-style as contributing factors to cancer. In the 1990s the news agenda was reinvigorated by renewed attention to tobacco and the discovery of new concerns such as electromagnetic fields.

The volume of coverage proved as variable as the topical focus. Figure 5.1 shows the ups and downs of cancer coverage during the

Figure 5.1 *Amount of Coverage Over Time*

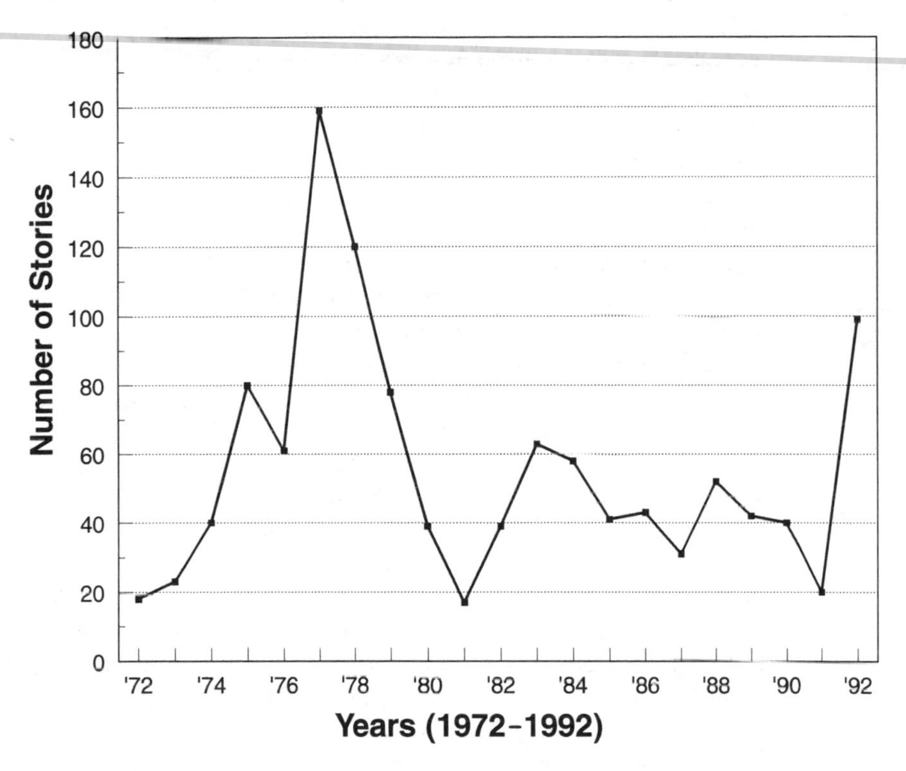

two decades reviewed by our study. It charts the steady rise in news volume throughout the 1970s. By the time the coverage peaked in 1977, it had risen nearly tenfold in only five years. The coverage then declined as precipitously as it had grown, bottoming out in 1981 at about the same level as a decade before.

Thereupon the amount of media attention to environmental cancer suddenly began to grow once again, until reaching a more sustainable level that remained fairly steady the rest of the decade. But just as it seemed that the saturation coverage of a decade earlier was a unique aberration, cancer news during the final year of the study leaped to levels not seen since the 1970s.

These sharp discontinuities in the coverage suggest the usefulness of a more detailed summary of the ever-shifting cancer story. The starting date for our analysis came soon after President Nixon declared a national "war on cancer." We detected no media offensive, however; the sample contained only seventeen stories in 1972, led by coverage of the ban on DES (diethylstilbestrol) in animal feed. Two years later a ban on two pesticides (aldrin and dieldrin) garnered significant media attention. But much of the coverage in these early years focused on genetic factors and viruses as causes of cancer. For example, the media gave considerable coverage to the work of David Baltimore and his colleagues, who received a Nobel Prize for research on how viruses cause cancer.

The volume of cancer news rose steadily during the mid 1970s as the focus shifted toward health risks associated with widely marketed consumer products. In 1976 the Food and Drug Administration (FDA) recalled a large quantity of peanut butter after it was found to contain aflatoxin—a naturally occurring carcinogenic mold product. Later the same year the agency banned red dye number two. Congress passed the Toxic Substances Control Act (TSCA), expanding the authority of the Environmental Protection Agency (EPA) and laying the groundwork for the Superfund Act. The TSCA included a phase out of PCBs (polychlorinated biphenyls) over a three-year period.

Media coverage of cancer skyrocketed during 1977 and 1978, when several scientific and policy debates collided. Saccharin was the number-one topic in 1977, as the debate moved from laboratories to FDA and congressional hearings. An initial FDA ban on saccharin was blocked by Congress and eventually modified to include restrictions and warning labels on products containing it. Later in the year the Consumer Product Safety Commission (CPSC) announced a ban on the use of Tris as a fire retardant additive to baby clothes after it was found to cause cancer in animal tests.

Also from the consumer front came calls to add hair dyes and sprays to the list of products policed by the FDA. The first asbestos case was settled in 1977 for several million dollars, and the first lawsuits over the use of DES in humans were filed. (By this time preliminary evidence suggested a link between DES and certain cancers in the female offspring of women given the drug when they were pregnant.)

Throughout 1978 media attention remained largely focused on man-made or human-induced causes of cancer. There was more coverage of asbestos litigation, the EPA proposed standards on filtration of carcinogens out of drinking water, and the CPSC outlined new guidelines to eliminate or reduce carcinogens in consumer products. The National Cancer Institute and the National Institute of Environmental Health Safety estimated that 20 percent of future cancers would come from workplace exposure to chemicals, with asbestos leading the list.

In 1979 the nuclear accident at Three Mile Island not only raised fears of radiation-induced cancers among nearby residents but also led to more general debate over the potential health risks from other commercial reactors across the country. That same year the surgeon general issued a large-scale study that explicated the linkage of smoking with cancer and a host of other ailments.

The news agenda broadened during the early 1980s even as the amount of coverage diminished, with concerns over man-made chemicals, smoking, and genetic aspects of cancer all receiving attention. In 1980 the Department of Labor issued its first comprehensive rules for the exposure of workers to carcinogens on the job. A new surgeon general released a study in 1982 indicating that smoking may cause one-third of all cancers and included a warning about secondhand smoke. His report contained the longest list of cancers linked to smoking ever issued by the surgeon general's office.

The coverage rose again in 1983, when Times Beach, Missouri, became a household name after the discovery of dioxin contamination throughout the town. A new wave of cancer scares continued during 1984 as the EPA banned the use of EDB (ethylene dibromide) as a soil and grain fumigant and began to work out plans for dealing with the already tainted grain. What some dubbed "the great muffin scare" cleared grocery store shelves until the FDA issued a statement declaring that EDB was not a problem in cooked products.

That summer the discovery of dioxin on part of Fort A. P. Hill threatened a Boy Scout National Jamboree. Eventually the scouts moved to a different part of the base. On another front, scientists at the National Institutes of Health (NIH) discovered the gene that mutated and became cancerous in the lungs of a smoker, thus providing definitive support for the oncogene theory of cancer causation.

In 1986, attention turned to radiation as a cause of cancer. First, an EPA report estimated that one in eight homes was contaminated with radon. Later came the fire at the Soviet Union's nuclear reactor in Chernobyl. Numerous experts held forth with varying estimates on the number of cancers that would be seen in the affected areas, as well as the implications for nuclear-related cancer risks in the United States.

The EPA was the focus of attention in 1987. First a National Academy of Sciences report recommended that the agency adopt a uniform set of standards for pesticide residues on all foods. This widely reported document argued that the food supply was inadequately protected from carcinogens. Later the EPA downgraded its estimates of the carcinogenicity of dioxin to one-sixteenth of the original estimate but continued to label it as the most carcinogenic chemical ever tested.

In 1988 the National Academy of Sciences estimated that radon may cause thirteen thousand lung cancer deaths and noted that smoking may compound the risk. Smoking also drew prominent coverage when a New Jersey jury held a cigarette company liable for cancer linked to its products. (Two years later a federal court threw out the verdict.)

With a push from *60 Minutes* and numerous concerned celebrities, the pesticide Alar became a prominently covered cause of cancer in 1989. A proposed EPA ban on its use forced the phase-out of

Alar over an eighteen-month period. For its part, the FDA proposed to ban or limit the use of some powerful animal drugs after early tests showed them to be carcinogenic. Residues of the drugs could be found in meat and milk from treated animals.

In 1990 the government opened a new front in its lengthy battle against the health risks of tobacco. The EPA proposed declaring secondhand smoke a known carcinogen, and the Occupational Safety and Health Administration (OSHA) considered secondhand smoke to be a work hazard. This controversy continued to percolate through the media until it was trumped by charges of tobacco industry cover-ups of incriminating research findings.

In 1992 the coverage jumped to levels not seen since the 1970s. As was the case then, the spike in news stories was the result of several public controversies. Among the earlier health concerns that received new attention were environmental tobacco smoke, asbestos, and the disposal of nuclear waste. These were joined by new fears of electromagnetic radiation as a possible carcinogen. Spurred on both by litigation and by epidemiological studies from Sweden, a spate of stories focused on the risks that might be posed by household appliances, cellular phones, and power lines.

This brief survey only skims the surface of the cancer-related news covered by our study. But it illustrates the diversity of topics that appear, as well as the different news "beats" involved. Information about suspected carcinogens comes from the science desk, but also from reporters assigned to cover Congress, various regulatory agencies, the courts, and many other beats. Cancer coverage also comes from general assignment reporters filing stories from the scenes of industrial accidents, political protests, and other newsworthy events that may evoke public fears of environmental cancer. Thus, the study of cancer news begins with science journalism, but it doesn't end there. Ultimately it traces the journalistic profession's treatment of the complex and multifaceted story that underlies this simple matter of life and death.

What Causes Cancer?

Now let us turn to the regularities that underlie this diverse and changeable set of concerns. To provide an overview of the news agenda on environmental cancer, we noted every statement by a reporter or news source identifying a particular cause of cancer. We distinguished between straightforward statements about known cancer-causing agents and those that identified "suspected" or "probable" causes of cancer.

This distinction, however, rarely altered the relative visibility of the substances most frequently in the news. That is, the same substances that were mentioned most often as undisputed carcinogens also got the most attention as likely suspects. So we tallied the number of times each substance was cited overall, while retaining the above distinction within the overall listing.

We categorized substances according to the way they were presented in a news story, without presuming any prior or expert knowledge by the reader or viewer. For example, the same substance might be identified in various stories as a pesticide, a pollutant, or simply a man-made chemical. We adopted the story's own designation in order to reproduce the various sources of risk just as they were communicated to the audience.

Man-made chemicals attracted by far the most media attention, as figure 5.2 reveals. This general category included 498 references to known or suspected chemical carcinogens, 70 percent more than were directed toward all forms of tobacco use. Many of these references were to unspecified "man-made chemicals," although specific substances were sometimes targeted. For example, a *New York Times* (1978) story on the Love Canal controversy noted that "82 different (chemical) compounds, 11 of them suspected carcinogens, have begun percolating through the soil. . . . Residents say many in the neighborhood have died of rectal, breast and blood cancers."

Many of these discussions raised concerns about the carcinogenic potential of the thousands of chemicals in industrial use that are subject to little or no government-mandated testing and regulation. These include numerous industrial solvents, components of plastics manufacturing, and chemical wastes, whose widespread usage predated the current governmental regulatory apparatus. For example, a *New York Times* article about liver cancer among plastics workers quoted a chemical plant physician: "I think when we look we'll find aniosarcoma of the liver all over the world wherever vinyl chloride is used" (Brody, 1974a).

Figure 5.2 *What the Media Say Causes Cancer, 1972–1992*

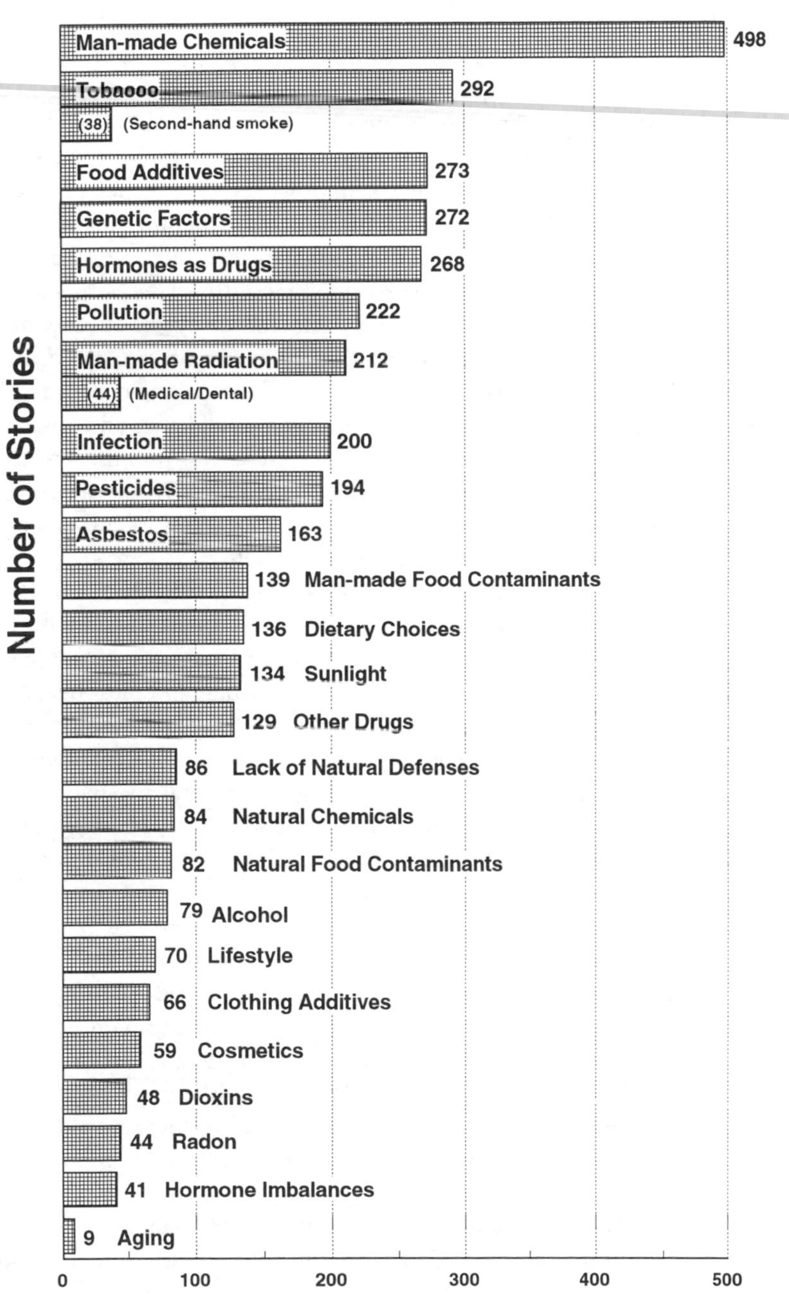

The paucity of test data and regulatory procedures to deal with these substances gave rise to suspicions about their potential for causing harm. These suspicions, which were often voiced by activist groups and other industry critics, frequently found their way into news stories. By the same token, however, the absence of regulatory battles sparked by government agencies sometimes deprived critics of opportunities to keep these substances under the media microscope long enough to create ongoing controversies.

Instead, newly developed chemicals or by-products, which are subject to more stringent regulatory oversight, generated many of the most memorable cancer scares of the past two decades. They include such items as food additives, pesticides, food contaminants such as residues from drugs that were given to animals, Tris and other clothing additives, and some cosmetics. We separated out the assertions that appeared about these categories of substances. The number of references to each is noted below.

Tobacco stands alone in its condemnation by the scientific community. But it finished a distant second in the national media coverage. The various forms of tobacco were identified as carcinogens 292 times in news reports. Of course, the earliest and most heavily publicized warnings by the surgeon general predated the time period that we studied. Nonetheless, considerable publicity attended the release of new studies by the surgeon general in 1979 and 1982, which expanded the list of cancers associated with smoking while increasing official estimates of the total proportion of all cancers that could be attributed to smoking.

There was also substantial coverage of additional periodic studies and official reports that gradually strengthened the case against smoking while broadening the list of cancers and the means of transmission beyond the original linkage of smoking with lung cancer. For example, later studies drew attention to cancers of the mouth and throat associated with tobacco use, while implicating snuff and chewing tobacco, as well as smoking. This coverage also included thirty-eight condemnations of secondhand smoke, the majority of which appeared during 1992 alone.

Our study excludes news about reported dangers of tobacco use aside from cancer, such as heart disease. It also ended before the recent flurry of antismoking initiatives and charges of tobacco company cover ups, which generated widespread publicity and refocused public attention on the dangers of smoking. Even had the study included this material, however, tobacco would have remained well behind chemicals in the cumulative tally of long-term coverage.[1]

Food additives were mentioned 273 times, nearly as frequently as all forms of tobacco. This category included substances such as artificial sweeteners, artificial food dyes, and synthetic food preservatives. Among the most prominently reported controversies referenced here are the 1976 FDA decision to ban red dye number two and the public furor surrounding the same agency's on-again, off-again effort to ban saccharine the following year.

Inheritance or genetic factors, including family histories, received about the same amount of attention as food additives (272 mentions). After generating some coverage in the early 1970s, genetic factors faded from view as public attention turned to the environmental cancer threats that are also our principal concern. They reappeared in the news in the latter 1980s, after advances in genetic research allowed scientists to identify the mutations of specific genes involved in the development of certain cancers.

For the most part, however, this category is made up of observations about how certain cancers tend to appear among members of the same family (such as breast cancer among women), suggesting the involvement of genetic factors. Since this is a relatively commonplace observation in general discussions of cancer causation, it is noteworthy how rarely genetic factors were targeted in the media, in comparison with the profusion of news about environmental factors. Genetics accounted for only 6 percent of all mentions of cancer causes.

The therapeutic use of reproductive-system hormones in human beings was cited 268 times. This category included the use of estrogen and progesterone as contraceptives (birth control pills) and the use of DES to prevent spontaneous abortions. (We excluded hormones given to livestock, as well as natural hormonal imbalances.)

Pollution was mentioned 222 times. This included human-induced air and water pollutants apart from pesticides, which were catalogued separately. Of course, this represents a small proportion of the references to all health hazards and dangers of environmental degradation associated with pollutants.

The same caveat applies to man-made radiation, whose carcinogenic properties were cited 212 times. This category mainly includes stories on nuclear accidents such as those at Three Mile Island and Chernobyl, along with other news about the dangers attendant to the commercial and military applications of nuclear power. For example, six years after the Chernobyl meltdown, an article in *Time* magazine (Jackson 1992) entitled "Nuclear Time Bombs" estimated that "10,000 deaths will result from fallout-induced cancers." The article charged that "[n]ot only is Chernobyl still a danger. So are many similar reactors, sunken submarines and radioactive waste dumps."

Also included within this overall category are a few references to medical and dental radiation such as X-rays, along with recent concerns about the carcinogenic potential of electromagnetic fields (although much of the ongoing EMF controversy is too recent to be included in our sample).

Viral infections were the only other cancer-causing agent to receive at least two hundred mentions. This category included statements linking the HIV virus, which causes AIDS, to otherwise rare cancers such as Kaposi's sarcoma. But many discussions of viral carcinogens predated the AIDS era, going back to heavy reportage of the 1975 Nobel Prize awarded to David Baltimore and his co-workers for their identification of the means by which viruses can cause cancer.

Pesticides were cited 194 times. This category included herbicides, insecticides, and rodenticides, whether directly applied to soil or crops, used as fumigants in crop storage, or found in food or water residues. The most commonly mentioned substance of this type was the fumigant EDB, whose presence in grain products led to a major cancer scare in 1984, which left consumers fearful of contracting cancer from packaged cereals and baked goods.

Asbestos completed the top ten list by receiving 163 mentions. Unlike most of our categories, this cancer risk involves a single substance rather than a number of related substances. Because of its widespread usage in insulation, asbestos has surfaced regularly in the news as a suspected carcinogen in the workplace, at home, and in public places. In 1974, for example, Jane Brody noted in the *New York Times* (1974b) that "thousands of workers have succumbed to asbestos-related cancers," adding that "new evidence raises questions . . . about the hazards the general public may face."

In addition to reports of scientific research, a combination of regulation and litigation has provided frequent cause for news organizations to consider the cancer risks associated with asbestos. Recent controversies have concerned the relative risks and benefits of removing this material from public buildings such as schools. But asbestos has been in the news repeatedly since the first of many court settlements took place in 1977.

Man-made food contaminants, apart from pesticides, were mentioned 139 times. They included growth hormones fed to animals, such as DES and bovine somatotropin (BST), fertilizer residues, and fluoride in the water supply. About the same amount of attention went to dietary choices, which received 136 mentions. Most of these concerned high-fat or low-fiber diets. A typical reference of this type was a *New York Times* (December 3, 1975) report that "[c]ancer of the colon . . . is related to an 'affluent' diet, particularly the excessive consumption of animal fats and the relatively low consumption of fibrous foods." This category excluded alcohol consumption and the ingestion of food additives and pesticide residues.

Sunlight was cited 134 times, usually as a cause of skin cancer or melanomas. Drugs apart from tobacco, alcohol, and reproductive hormones accounted for 129 mentions. This category included marijuana, hypertension drugs, and sleep aids. All other categories of suspected carcinogens received fewer than a hundred mentions apiece.

The lack of natural defenses, cited eighty-six times, refers to chemical deficiencies that reduce the body's natural ability to fight off natural cancer-causing agents. Natural chemicals (apart from the food contaminants listed below) were mentioned eighty-four times. This category refers mainly to general discussions of chemicals that are not identified as man-made.

Natural food contaminants were cited eighty-two times. The best known of these is aflatoxin, a fungus by-product found in peanut butter. Fear of aflatoxin contamination led to an FDA recall of peanut butter in 1976. A *Washington Post* article (Burros 1976) on the agency's action was headlined, "Peanut Butter: Is Nothing Sacred?" Alcohol consumption (including products such as mouthwash, which contain alcohol) was listed seventy-nine times.

Life-style factors not listed elsewhere accounted for seventy mentions. These included sedentary habits and high stress levels. Completing the top twenty causes of cancer were clothing additives, which received sixty-six mentions. These included dyes, the fire retardant Tris (which was banned in 1977), and chemicals used in clothing production.

Cosmetics, which were cited fifty-nine times, included various toiletries and personal-care items such as hair dryers, hair sprays, and tanning agents in suntan lotions. A flurry of publicity attended the release of studies suggesting that many chemicals in hair dyes and sprays were possible carcinogens. As a result, these products were brought under the FDA's jurisdiction.

Dioxins and furans were mentioned forty-eight times as cancer agents, most often in connection with Agent Orange and contaminated soil in Times Beach, Missouri. The relatively low ranking of dioxin on our list is a reminder that heavy coverage of a particular health scare does not necessarily translate into a high profile for a carcinogen over a twenty-year period. In addition, many dioxin-related stories emphasized its dangers in general terms as a health hazard, without specifically identifying it as a cancer agent. This was the case, for example, in news reports about the 1984 Boy Scout National Jamboree, which was scheduled to take place on a dioxin-contaminated site.

The carcinogenic potential of radon emissions was noted forty-four times. In this instance, the apparent paucity of references to a controversial substance reflects the relatively recent discovery of its potential dangers. Widespread public concern about radon contamination dates from 1986, when the EPA established safety levels and began a public education effort.

Finally, hormonal imbalances were cited in relation to breast cancer and reproductive system cancers forty-one times; this category excludes therapeutic uses of reproductive-system hormones. No other category of cancer agents received as many as ten mentions. Aging was referenced nine times in connection with the tendency of cancer incidence to increase with age.

The Media versus the Experts

This news agenda stands in stark contrast to the concerns voiced by cancer researchers, whose views were described in chapter 4. Over the past two decades, the media have paid far more attention to man-made chemicals than to any other category of carcinogen, including tobacco. They have also given heavy coverage to the cancer risks of food additives, pollution, pesticides, radiation, and hormone treatments. News stories targeted each of these substances as cancer agents more often than diet, sunlight, or asbestos, which the scientists regarded as more potent cancer threats. Indeed, media reports named food additives as carcinogens as often as diet and sunlight combined.

Thus, the environmental factors stressed by the cancer experts differed sharply from the dangers that we read and hear about most frequently. Of course, it might be objected that the data on news coverage were accumulated across two decades, and cancer researchers themselves might have answered quite differently earlier in our study period. For several reasons, however, it seems unlikely that this factor would significantly alter the pattern of survey results reported here.

First, as we noted in chapter 4, scientists' rankings of risk factors proved remarkably stable over the eight-year interlude between the 1985 pilot study and the final 1993 survey. Second, both these surveys produced results that accord closely with the Doll-Peto synthesis of the field published in 1981. Their emphasis on tobacco and diet, as well as their deemphasis of man-made chemicals, additives, pollutants, and the like, reflected a comprehensive survey of the existing research literature.

In other words, the Doll-Peto classification consolidated the already existing knowledge of the research community, rather than representing any sort of revision to mainstream scientific thinking. However much these results challenged the media-based conventional wisdom on environmental cancer, they simply consolidated the view of cancer risk already prevalent among cancer researchers.

Finally, as an additional check, we compared the 1993 survey ratings with the content analysis data from 1992, the final year in the sample. If the media were keeping abreast of expert opinion, there should have been a closer match between current scientific opinion (as measured by the survey data) and the current news coverage than there was with the cumulative long-term media portrait. Unfortunately, the most recent cancer news agenda still had more in common with previous patterns of cancer coverage than with current expert opinion.

In 1992 therapeutic uses of reproductive-system hormones received the greatest attention among all environmental causes of cancer. Despite the growing controversy over secondhand smoke, tobacco once again finished second. Man-made chemicals placed third on the list, followed by nuclear power. The only shift in the direction of scientific opinion came in the rise of sunlight to fifth place on the roster of cancer agents noted most by the media.

Dietary choices, however, again failed to break into the top ten, garnering only about half as many mentions as chemicals and one-third as many as hormones. Once again, diet finished behind such longtime objects of media attention as pesticides and artificial food contaminants, in addition to hormones, chemicals, and radiation.

Of course, any such one-year listing may be strongly affected by current events and other transient factors, such as the controversy over estrogen replacement therapies that made news in 1992. That is why we examined the media coverage over a much longer time period. But these results demonstrate that the differences between scientific opinion and media coverage cannot be dismissed as artifacts of the timing of the survey.

By acknowledging the role of transient events in skewing news coverage of public health issues, we must address a more serious objection to our interpretation of these findings. Can journalists be held responsible for the deviation of news from expert opinion on a scientific issue, or is this simply a reflection of the real-world events that they dutifully report? The divergence between the way scientists assess cancer risks and the way the media communicate them clearly represents a problem for journalism. It is less clear precisely how—or even whether—it represents a failure of journalism.

We recognize that the news is not a running literature review of research findings. Among other things, it is a chronicle of significant public events and other matters of popular interest. Journalists are far more attuned to the world of public affairs than to that of research laboratories. As a result, reporting on scientific issues will typically be skewed toward stories that emphasize individual personality, public policy, political conflict, and immediate rather than long-term consequences. In recent years journalists have also tended to emphasize the "system" rather than individual responsibility.

As we noted earlier, the cancer risk "story" is actually a compilation of numerous and only partly overlapping stories. The subject matter involves various combinations of litigation, legislation, and regulation, as well as scientific research. The concerns and language of lawyers, politicians, and interest groups are often very different from those of scientists. Journalists can hardly be blamed for reporting the realities of public debate, even if public discourse diverges from scientific discourse.

This is a familiar argument in defense of divergent representations of reality by journalists who cover a story and individuals who are more intimately involved or familiar with it by virtue of their professional activities. Indeed, some form of this argument is raised quite frequently by journalists responding to charges of biased, sensational, or superficial reporting of various subject areas in which some form of expert knowledge is involved, ranging from economics to risk assessment.

What is noteworthy about the media portrayal of cancer causation, however, is that it was rarely linked to players in the science policy process. Instead, the vast majority of these statements were either attributed to scientific experts or were presented by reporters without attribution. Nearly half (43 percent) of all mentions of the environmental causes of cancer took the form of unattributed statements by reporters, which were presented as direct representations of reality rather than assertions by sources.

Another 29 percent involved direct or indirect quotations from specific expert sources or were variously attributed to unidentified "experts," "scientists," or "researchers." That left only 28 percent of all statements about cancer causation to be apportioned among newsmakers and other interested parties, such as government officials, corporate spokespersons, environmentalists or other "public interest" groups, and members of the general public.

Table 5.1
Cancer Causes Cited in News Stories, 1972–1992

	Overall Rank	Reporters	Expert Sources	Other Sources
Manmade chemicals	1	1	1	1
Tobacco	2	8	3	2
Food additives	3	2	4T	7
Inheritance	4	3	2	19
Hormones as drugs	5	4	4T	10
Pollution	6	5T	7	4
Manmade radiation	7	11	11	3
Infection	8	5T	6	15
Pesticides	9	9	8	5
Asbestos	10	5T	12	11
Manmade food contaminants	11	12	10	9
Dietary choices	12	14	9	13
Sunlight	13	10	16	6
Other drugs	14	13	13T	8

Lists include all substances cited at least 100 times; T indicates a tie-in number of mentions

Moreover, the relative prominence accorded to various cancer risks remains very similar regardless of the type of source to whom this information is attributed. In table 5.1, the most frequently cited carcinogens (more than one hundred media mentions) are listed according to the source who presented this information. The list of substances identified by "expert" sources closely resembles that of journalists speaking or writing on their own authority.

For example, reporters themselves, as well as the "expert" sources they cited, named man-made chemicals more frequently than any other carcinogen. Apart from tobacco, which reporters mentioned somewhat less frequently, the top ten lists for both groups look remarkably similar. The remaining sources were more varied in their portrayals of cancer agents, focusing more frequently than either reporters or experts on radiation and sunlight, less frequently on infection, hormones, and inheritance.

This suggests that the divergence of media portrayals from scientific opinion reflects something more than the words and deeds of newsmakers. It reflects journalistic judgments about which sources are reliable, how representative various experts are, and whether their assessments are consensual or controversial in their field of expertise.

After all, the media have not reported uncritically the perspectives of those who oppose fluoridation or Darwinian theory, even when their challenges are couched in scientific terms and presented by individuals with academic credentials, because their opinions are presumed to contradict the expert consensus on these matters.

Yet the scientific consensus on environmental cancer risk that emerges from survey data differs dramatically from the risks cited by news sources who are specifically identified as experts, as well as those who are not. That is, a true cross-section of cancer researchers would give much greater weight to the cancer risks associated with diet and sunlight and much less weight to the risks of man-made chemicals, food additives, pollution, and pesticides than the nonrandom "sample" of experts whose opinions are cited in news stories.

Reporting Scientific Controversies

Our interpretation of journalism's active (if unintentional) role in distorting public perceptions of environmental cancer is bolstered by the way the media reported on substantive debates over cancer causation. So far we have been dealing only with the relative amounts of attention paid to various putative carcinogens. Now we turn to an arena in which one could reasonably expect journalists to "get it right" without being constrained by legitimate news values that might inadvertently skew the overall picture of reality. This involves the reporting of self-contained controversies on which positions pro and con are set forth and argued in the public fashion that characterizes scientific debate.

We examined the coverage of several highly visible scientific controversies involving the source, extent, and detection of environmental cancer risks. These include the following questions: Do we face a cancer epidemic in this country? Are carcinogens unsafe at any dose? Can human cancer risks be inferred from animal tests? And does a single finding of carcinogenicity constitute an appropriate standard of proof for banning a substance from human use?

Figure 5.3 *Cancer Controversies: Scientists vs. Media*

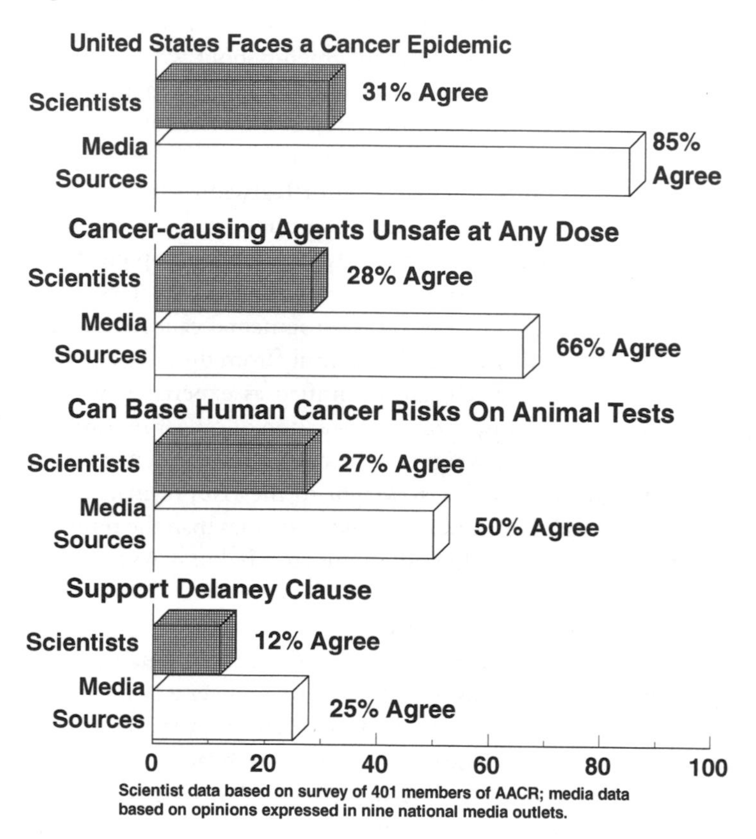

Scientist data based on survey of 401 members of AACR; media data based on opinions expressed in nine national media outlets.

These are all controversies on which we gathered opinions in our survey of cancer researchers. So in each instance we shall ask whether the media coverage reflected the dominant view among experts, provided a balance between opposing positions regardless of where most expert opinion lies, or presented perspectives that usually conflicted with the expert consensus. The results appear in figure 5.3.

Is cancer so prevalent in contemporary American society that it can be said to have reached epidemic proportions? The notion of a "cancer epidemic" has surfaced repeatedly in public discussion of environmental health risks over the past quarter-century. It often appears in conjunction with concerns that the products of industrial activity, such as pollution and artificial chemicals, help to account for the observed rise in cancer rates over the past several decades. Alternatively, fears may be expressed that widespread past exposure to substances newly identified as carcinogens has created the time bomb of a future epidemic. So it may not seem surprising that sources queried on this question have usually agreed with this proposition. Five out of every six opinions that were printed or broadcast supported the argument that America faces some sort of cancer epidemic. Typical of concern over a future epidemic was a 1984 *New York Times* article citing a team of Johns Hopkins University scientists who "contend that 'epidemic increases' in multiple myeloma, brain and lung cancers" represent a warning "that cancer deaths from chemicals could surge in the future."

Yet the media consensus lies on the opposite side of the spectrum from most cancer researchers. In our survey, two out of three scientists (67 percent) rejected the notion of a cancer epidemic, and only 31 percent assented. In answer to a separate question, when asked whether rising cancer rates mainly reflect the results of industrial activity or the combined effects of tobacco and aging, two out of three scientists (65 percent) selected the latter, compared to only 15 percent who blamed industrial activity. Another 15 percent responded that both were equally to blame, and 5 percent were unsure.

Are cancer-causing agents unsafe at any dose? As a technical question, this involves the determination of tolerance levels for carcinogenic substances. A watchword of toxicology is that the dose determines the poison, and most researchers apply this principle to suspected cancer agents. Respondents in our survey of experts disputed the assertion that carcinogens are unsafe regardless of the dose, by a margin of more than two to one (64 percent to 28 percent, with the rest unsure).

In the media, however, these proportions are inverted, with two out of every three sources (66 percent) arguing that humans have zero tolerance for cancer agents. For example, Dr. Sidney Wolfe told the *New York Times,* "The argument that testing can find such substances in smaller and smaller measures is true. But we're also finding out that smaller and smaller amounts of such substances are linked to cancer" (De Witt 1980).

In fact, among journalists' sources who assessed current legally mandated tolerance levels, a majority argued that they were too high (i.e., they permit a greater level of exposure than humans can tolerate). Tolerance levels were criticized as being too high three times as often as they were called too low. (The exact margin was 54 percent to 18 percent, with the other 28 percent of those quoted finding current levels acceptable.) During the debate over how much aflatoxin should be permitted in peanut butter, the *Washington Post* (Burros 1976) noted a consumer organization's argument that "since it is possible to have aflatoxin-free foods the tolerance level should be set to zero."

Can the results of animal studies of suspected carcinogens be extrapolated to humans, in order to assess the health risks associated with specific substances? This is a standard procedure for establishing carcinogenicity in accordance with federal regulations. This approach has long been controversial, however, since it involves giving very high doses of substances to animals even though humans are exposed to far lower doses of the same substances.

Media coverage of this controversy has been almost perfectly balanced, with 50 percent of sources affirming this approach and 49 percent denying its acceptability. For example, the *Washington Post* (Burros 1976) quoted Dr. Sidney Wolfe's argument that "[e]ven though direct evidence is unobtainable . . . virtually all substances known to cause cancer in humans, also cause cancer in experimental animals." The opposing argument was presented by EPA administrator John Moore, who told the *New York Times* (Shabecoff 1988) that "in some cases what happened to a mouse has no relationship to human cancer."

This, however, is an instance in which numerical balance of opposing viewpoints does not correspond to the actual split in expert opinion. Only 27 percent of the cancer researchers endorsed the current practice of assessing human cancer risks by giving animals what is termed the "maximum tolerable dose" of suspected cancer-causing agents. More than double that number (63 percent) disagreed, with the remainder unsure. So majorities of scientists and media sources once again came down on opposite sides of the issue, though by a slimmer margin this time.

The media coverage roughly paralleled expert opinion on only one of the four issues on which we compared them systematically. This involved the principle embodied in the Delaney clause, which held that chemicals and additives must be banned from food if they ever are shown to cause cancer in any species. Scientists overwhelmingly reject this notion, by 85 percent to only 12 percent (with 4 percent unsure) among those surveyed. This time the media sources were not far behind in their rejection rate, with three out of four (75 percent) calling for the repeal of the Delaney clause. That is still a higher rate of agreement than we found among scientists, of course, but at least the bulk of opinion was on the same side for once.

The results of the two surveys are not quite as comparable on this question as they are in the previous three controversies, since the sources were frequently speaking to an issue of regulatory policy that codified a risk assessment standard, rather than to the scientific debate underlying that standard. A 1980 *New York Times* story (De Witt 1980), for example, cited the Reagan administration official James C. Miller III's argument that "the Delaney clause ought to be replaced with a 'risk benefit' standard that takes into account the economic consequences of eliminating certain chemicals from the marketplace." In the same article, Dr. Sidney Wolfe replied, "Food additives aren't really needed and are a small thing to forgo to prevent society's exposure to cancer."

Thus expert opinion frequently differed from news reports on scientific controversies, just as it differed from reports on the causes of cancer. Relative to the actual views of cancer researchers, the media covered scientific controversies in a manner that overestimated the extent of the cancer threat and advocated more stringent standards for evaluating health risks associated with suspected carcinogens than most experts believe is necessary.

This systematic pattern of discrepancies between actual expert opinion and the popular media portrayal of cancer risks might be interpreted benignly as erring on the side of caution. A less charitable reading would accuse the media of creating unnecessary cancer scares and widespread "cancerphobia" among the public. Either way, the results raise troubling questions for science journalism.

It might well be argued that media coverage should give equal voice to both sides of scientific controversies in the interest of fairness and balance, just as in political coverage. But it is difficult to argue that the media serve the public by giving credence to positions that are rejected by most experts in the field, or by deflecting public attention from more important health problems onto less important ones.

Thus, the trouble is not only that the news per se differs from expert opinion, leaving the public poorly informed about cancer risk; the deeper problem is that the public is being misinformed about the nature of expert opinion on cancer risk. The latter is not only a more serious impediment to an informed public, it is also a more direct indictment of the journalistic profession.

HOW THE EXPERTS RATE THE MEDIA

If this criticism seems unduly harsh, consider how closely it is echoed by cancer experts themselves. The disjunctions between expert opinion and media attention uncovered by our research proved a close match with these scientists' own criticisms of the coverage. As part of the survey, we asked them to rate media coverage of various risk factors in terms of whether the reported cancer risk of each substance was overstated, understated, or stated fairly. The results appear in figure 5.4.

Majorities of the cancer researchers held that the media overstate the risks associated with nuclear power (61 percent), pollution (54 percent), and food additives (53 percent). Pluralities felt the same

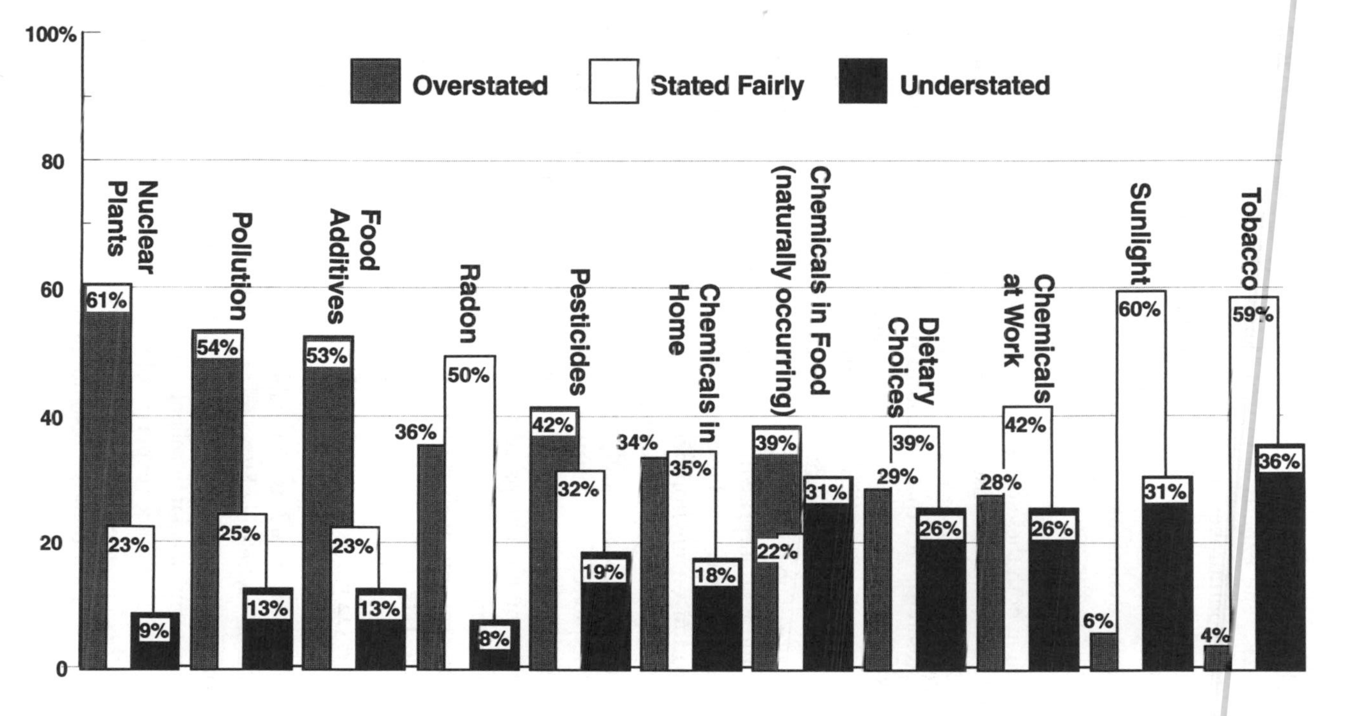

Figure 5.4 *How Scientists Rate Media Portrayals of Cancer Risks*

way about coverage of chemicals occurring naturally in foods and pesticides. The experts who complained that reported risks were overstated outnumbered those who complained that they were understated by margins of six to one on nuclear power, four to one on pollution, food additives, and radon, and two to one on pesticides and chemicals present in American households.

Only in the case of sunlight and tobacco were scientists more likely to rate the reported risk as understated than as overstated (though in both these cases, majorities felt that the risks were stated fairly). Thus, the scientists' own assessments of the biases in journalists' coverage of environmental cancer closely correspond to the actual discrepancies that we found between the expert risk assessments and the national media's coverage.

Consistent with these criticisms, many cancer experts expressed doubt about the reliability of national media accounts of their field. Their vote of no confidence in the media is recorded in figure 5.5. It may come as no surprise that only one out of sixteen scientists in our sample (6 percent) rated television news as a highly reliable source of information on environmental cancer, while nearly ten times as many (55 percent) rated television newscasts as unreliable sources. But the ratings were hardly any better for the weekly news magazines, despite their advantages of running longer stories on more leisurely deadlines. Only 9 percent of cancer experts rated news magazines highly, compared to 49 percent who gave them low reliability ratings.

Most striking of all, though, was the lack of scientific respect for the *New York Times,* which is renowned among journalists for its award-winning weekly science section. Fewer than one in four researchers (22 percent) rated the *Times* as highly reliable in its cancer coverage. In fact, the proportion of cancer experts who rated America's paper of record as unreliable exceeded the proportion who found it highly reliable by 30 to 22 percent. As a baseline for comparison, 54 percent rated *Scientific American* as highly reliable, while only 8 percent called it unreliable.

Presumably cancer experts hold the national media in such low esteem because the treatment of environmental cancer accords so poorly with their own understanding of their field. But how could

Figure 5.5 *Scientists Who Trust Cancer Information From:*

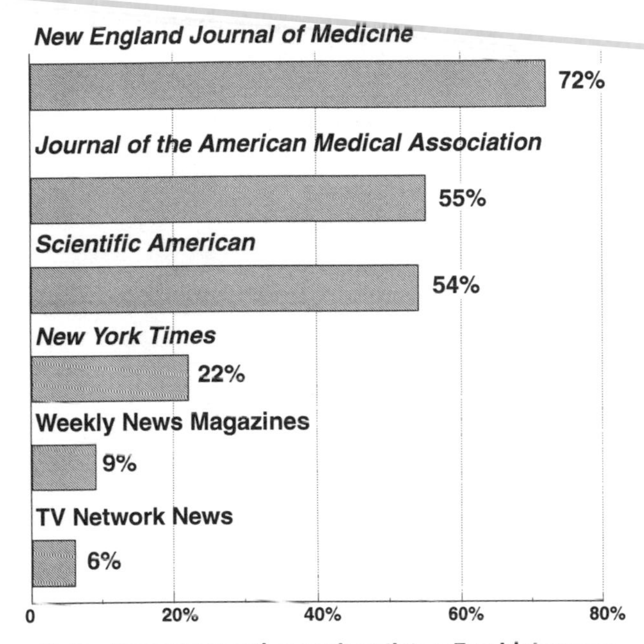

Note: Percentage rating each outlet as 7 or higher on a 0–10 scale of reliability.

Table 5.2
Scientists' Ratings of Level of Confidence in Expertise on Environmental Cancer

Individual Rated	High	Medium	Low	DK
Bruce Ames	67	19	6	8
Richard Peto	57	15	3	24
Richard Doll	53	8	2	37
Samuel Broder	49	18	7	26
Robert Weinberg	47	18	11	24
Joseph Fraumeni	40	15	4	41
Irving Selikoff	30	15	4	41
John Higgenson	29	12	4	56
Sidney Wolfe	24	15	11	50
Samuel Epstein	24	20	17	40

Ratings on a 0–10 scale: low = 0 to 3, medium = 4 to 6, and high = 7 to 10, DK = don't know.

such a disparity come about? One likely reason is that many "expert" sources are not highly regarded by the scientific community. When we asked these researchers how much confidence they had in the expertise of some well-known individuals in their field, they gave their lowest ratings to some whose names show up most frequently in the news.

As table 5.2 shows, the lowest ratings on reputation went to the two scientists on the list who have been prominent critics of suspected man-made carcinogens. They are the medical researchers Sidney Wolfe and Samuel Epstein. Dr. Wolfe attained his high public profile as longtime head of the Ralph Nader–affiliated Public Citizen Health Research Group. Dr. Epstein's Cancer Prevention Coalition frequently criticizes industrial carcinogens.

On the basis of our content analysis, Wolfe and Epstein were among the ten most frequently quoted scientists in the news sample. Yet they each received a high confidence rating from fewer than one in four cancer researchers—only 24 percent of those surveyed.

By contrast, the highest confidence rating—67 percent of those surveyed—went to the Berkeley biochemist Bruce Ames, whose reservations about the carcinogenic effects of synthetic substances seem closer to the mainstream among cancer researchers. Ames also reaches a national audience as the fourth-most-quoted cancer expert in the sample, following Wolfe, Irving Selikoff, and the former Stanford University president Donald Kennedy.

Close behind Dr. Ames in ratings on reputation were Richard Peto and Richard Doll, whose conclusions about cancer causation, described in chapter 3, are quite similar to those of Ames. In fact, based on mean ratings that exclude respondents not rating each individual, Doll finished slightly ahead of Ames atop the list, while Epstein fell slightly behind Wolfe at the very bottom.

THE BATTLE FOR THE HIGH GROUND

The high media profiles of the three scientists whose work received the lowest ratings from their peers suggests the direction in which cancer news is frequently tilted. As we have seen, the media's treatment of environmental cancer has little in common with the views of most cancer researchers. By contrast, the news agenda on cancer risk accords rather better with the assessments of environmental activists. In general, the media have paid far more attention to the cancer risks that preoccupy environmental activists than to those that concern scientific experts.

Recall that the environmental leaders rated such factors as pollution, food additives, nuclear power, and the pesticides DDT and EDB as significantly greater cancer risks than did scientific experts. All these factors are also on the short list of the most newsworthy cancer risks. Further, the activists were far less likely than the scientists to believe that the media are overstating the risks associated with these factors.

Specifically, when environmental leaders were asked to rate the media's treatment of several risk factors, their responses were almost invariably more muted than those of the cancer researchers. The comparisons are summarized in table 5.3. More than six out of ten scientists criticized the overstated reporting of nuclear risk, compared to just over one out of six activists. Where a majority of cancer

Table 5.3
Scientists' and Activists' Ratings of News Reports of Cancer Risks

	Scientists	Activists
Nuclear plants	61%	18%
Pollution	54	8
Food additives	53	9
Pesticides	42	4
Radon	36	12
Household chemicals	34	8
Smoking	4	2

experts perceived the media as overstating the risks of pollution and food additives, the percentage of like-minded environmentalists was in the single digits. And the experts were more than ten times as likely as the activists to criticize the media's portrayal of pesticides as a cancer risk.

One obvious reason for this discrepancy is the congruence of interest between the media and the environmental movement. Environmental groups are in the business of making news, just as journalists are in the business of reporting it. Thus, the *New York Times* science editor Nicholas Wade affirms that the press often functions as a "passive conduit" for environmental critics: "Often we're just doing our duty in following the activism of environmentalists, who make an issue of radon in houses or abandoned Superfund sites. Then it gets taken up in Congress and we have to cover it" (Kurtz 1993).

Note that Wade's statement is not a *mea culpa,* much less a confession of personal advocacy. Instead, he presents it as a recognition of the structural biases of a profession most attuned to the aspects of any event or development that impinge on public affairs, governmental activities, and social conflicts. Such emphases carry unintended consequences for journalists who are "doing their duty" by covering the activities of newsmakers. But journalists are not to blame for this situation; it is built into their job of reporting on public affairs.

Obviously, there is some truth in this defense of an inevitable media tilt toward the perspectives of social activists. To the extent that the multifaceted cancer "story" intersects with the activities of environmental groups, the coverage inevitably reflects the lawsuits they bring, the studies they release, and the political pressures they exert. But this argument is only half the story. Indeed, it represents one side of an ongoing debate among journalists themselves over their role in this process.

For some opponents of Wade's position, such pleas of professional necessity become evasions of journalistic responsibility. Among these critics is Hodding Carter III, whose experience spans the spectrum from newspaper publisher and television producer to columnist and State Department spokesman. Mr. Carter has ridiculed what he calls "the media's version of the Nuremberg defense . . . : 'We only report (do) what others tell us (order us) to do.' " Attacking his profession's contribution to the Alar scare, he concluded, "And that's the media's problem as it goes about manipulating and being manipulated to the end, intended or accidental, in defining the all-important high ground of public perception. . . . Our own worst habits are responsible, which means the place to begin is within the newsroom" (Carter 1989).

However outlandish some journalists might find the Nuremberg analogy, few commentators would contest that the media manipulate as well as being manipulated. The media should at least be considered as codependents in the addiction to recasting reality in altered form as news. Surely the term "passive conduit" is not a self-evident description of the media's current role in American society.

Journalists do not simply play their assigned role in the political narrative. They help to write the script, as active participants in the process of creating public perceptions and shaping public discourse. As the sociologist Dorothy Nelkin concluded from her study of biomedical reporting, "Through their selection . . . they set the agenda for public policy. . . . Through their style of presentation they lay the foundation for public attitudes and actions" (Nelkin 1985, p. 643).

Thus, interest groups regularly seek to make news in order to further their political objectives. By drawing attention to "their" issues and framing these issues in terms of their choosing, they increase the likelihood of generating public concern, political support, and favorable regulatory and judicial outcomes. The media so pervade this process that making news is making policy by other means. This is particularly true for an issue like environmental health risk, which engages primal fears of physical safety and well-being.

Indeed, the Alar episode illustrates the degree to which journalism can become enmeshed in the politics of social activism. The Natural Resources Defense Council turned to the media after being rebuffed by both the courts and the Congress in its efforts to ban the pesticide Alar from the food supply. The NRDC used a public relations firm along with their own media savvy to turn their much-disputed Alar study into a major news event. They accomplished this by using an exclusive *60 Minutes* report to launch a coordinated publicity campaign that included a television commercial and congressional testimony featuring the actress Meryl Streep (Weisskopf 1990a).

The resulting saturation media coverage whipped up a public frenzy that forced Alar off the market and cost the apple industry over $100 million in lost sales. This rendered irrelevant Surgeon General C. Everett Koop's judgment that "Alar-treated apple products posed no hazards to the health of children or adults." After all, forty million viewers had already heard Ed Bradley declare Alar to be "the most potent cancer-causing agent in our food supply" (Shaw 1994c).[2]

In fact, the very visibility of this highly successful end run around the governmental and scientific establishments proved to be its only drawback, by creating skepticism about such future maneuvers. The news management was so overt (including a self-congratulatory memo by the NRDC's public relations firm, which found its way onto the op-ed page of the *Wall Street Journal*) that it generated a backlash among many journalists, who were appalled by the media's eager participation in a rush to judgment.[3] A *Washington Post* editorial complained that "a complicated scientific issue" was decided "by a frightened public acting on incomplete and often erroneous press reports." A *Los Angeles Times* environmental reporter (Shaw 1994c) condemned the episode as "completely alarmist. . . . A lot of media people were suckered in." And Hodding

Table 5.4
Scientists' Ratings of Level of Confidence in Institutional Expertise on Cancer

	High	Medium	Low	DK
National Cancer Institute	92%	7%	0%	1%
National Institute of Health	87	9	2	2
Centers for Disease Control	70	21	6	3
World Health Organization	64	28	6	2
Center for Science in the Public Interest	15	22	12	51
Environmental Defense Fund	16	29	21	33
Tobacco Institute	5	14	73	8

Ratings on a 0–10 scale: low = 0 to 3, medium = 4 to 6, and high = 7 to 10, DK = don't know.

Carter (1989) took the occasion to hurl a more general criticism at his colleagues: "The Alar situation is a case study in the way today's journalism often makes truth the first casualty of its most treasured conventions."

The last point is especially noteworthy. For all the notoriety that attended the Alar story, it was just a particularly successful instance of techniques that are used routinely by such organizations. For example, the *Washington Post* media critic Howard Kurtz has described how the Center for Science in the Public Interest (CSPI) created a public furor over saturated fat in movie theater popcorn. The CSPI "carefully stage-managed the news," Kurtz wrote (1994), by first sending an embargoed copy of its popcorn study to numerous reporters to build interest, then leaking the findings to a select few, and holding a news conference that featured "colorful visuals" and "tantalizing sound bites."

Such skillful uses of public relations techniques are better suited to impressing journalists than scientists. Thus, the cancer researchers we surveyed showed surprisingly little regard for information from environmental groups, just as they had downgraded the expertise of scientists who made their names as environmental activists. See table 5.4, which shows how the experts rated a list of organizations as sources of reliable information on environmental cancer.

We asked our sample of scientists to evaluate two randomly chosen influential mainstream environmental groups to which the media tends to give respectful attention. Both fared poorly. Only one in six scientists (16 percent) rated the Environmental Defense Fund as highly reliable. Even fewer (15 percent) gave a high reliability rating to the aforementioned Center for Science in the Public Interest. By comparison, the National Cancer Institute was rated as highly reliable by 92 percent; at the other end of the spectrum, the Tobacco Institute was rated highly by 5 percent.

As these cases also illustrate, however, journalists have numerous incentives to publicize the activities of such groups. Environmental organizations serve as enthusiastic sources of background information, provide colorful quotes or sound bites, and create breaking news that promotes social controversy, political conflict, and journalistic careers. If journalists believe that a particular group's agenda puts it on the side of the angels, so much the better. For journalists and activists alike, advancing the public interest then goes hand in hand with advancing one's personal and professional interests.

Certainly, major media journalists tend to sympathize personally with the aims of these organizations. As a group, journalists share the socially liberal perspective that we found to be typical of the environmental movement's leadership. Among both groups, surveys locate environmentalism as one component of a cluster of attitudes that includes suspicion of business and the military and support for feminism, affirmative action, gay rights, and abortion rights.

These are all issues associated with the social liberalism commonly found among contemporary upper-status professionals in the United States. Researchers for a *Los Angeles Times* survey of the profession (Schneider and Lewis 1985, p. 7) concluded that most newspaper journalists could be described as "yuppie liberals." Our own survey of national media journalists (Lichter, Rothman, and Lichter 1986) used the term "liberal cosmopolitans" to designate the predominant outlook among these individuals.

In fact, cross-national studies (Inglehart 1990) show that environmentalism around the world is associated with a value orientation that emphasizes "self-expression and the quality of life" over "economic and physical security." Researchers find that these "subjective predispositions" are as important as "objective" factors (such as pollution levels) in predicting environmental advocacy and activism (Inglehart 1995, p. 57). Among people who share this sensibility, it is easy to equate the environmentalist agenda with the public interest.

Indeed, journalists can be quite open about where they stand in reporting on environmental controversies. In 1989 *Time* magazine's science editor, Charles Alexander, proudly told an environmental conference, "I would freely admit that on this issue we have crossed the boundary from news reporting into advocacy." NBC's Andrea Mitchell added that "clearly the networks have made that decision now, where you'd have to call it advocacy."[4]

These individual comments are born out by systematic survey data. In one study, 240 randomly selected national media journalists were asked to name a reliable source on environmental problems. Sixty-nine percent mentioned environmental activist groups, more than ten times as many as the 6 percent who cited scientific journals such as *Science* or *Scientific American*. In the same survey, more than four out of five journalists (81 percent) rejected the notion that America's environmental problems are overstated (Lichter, Rothman, and Lichter 1986, pp. 29, 57).

The bottom line is that any number of groups and interests regularly seek to use the media to advance their interests. Some are noticeably more successful than others. A critical variable in their success is the receptivity of journalists to their message. It does not hurt a group's chances of making news for journalists to agree with its social mission and share the worldview of its leadership. Journalistic resentment at being manipulated is activated more readily by the efforts of adversaries than by those of allies.

Nor should it come as any shock that journalists would tend to emphasize perspectives with which they empathize. This does not even require conscious advocacy; it may simply reflect the influence of "news judgment" in shaping story and source selection and the interpretation of ambiguous information. Professional training and adherence to norms of fairness and balance are supposed to prevent personal biases from spilling over into news coverage, whether intentionally or unconsciously. But these measures can be effective only when journalists define an issue as having two legitimate sides. And this is not the mind-set that they often bring to issues of environmental health risk.

This problem was noted by David Shaw, the *Los Angeles Times*'s Pulitzer Prize–winning media critic, who recently documented the tendency of journalists to "write from a pro-environmental standpoint and to make common cause with environmental activists." Shaw attributed this predilection to "the emotional identification that many reporters felt with the environmental movement. To them, 'saving the earth' was not the sort of partisan political or ideological issue they had been trained to avoid taking sides on; it was a nonpartisan effort to make life better for everyone" (Shaw 1994b).

Whether the symbiotic relationship between journalists and activists serves the public interest depends upon one's assessment of the causes being advanced. In the case of science policy issues, however, it clearly skews the news agenda toward scientific perspectives that are advanced either through interest groups or by individual scientists interested in using their work to promote social change.

This relationship also undermines the position of journalists who argue that the media should be credited rather than criticized for emphasizing health risks in the interest of social reform. A prominent adherent of this perspective is the *New York Times* editor Howell Raines, who asserts, "Risk is not just about the statistical threat to health and life. Risk is also about what kind of society you want to have." Unfortunately, this conception of risk tends to sacrifice accurate information to the shared enthusiasms of journalists and activists who want to have the same kind of society (Shaw 1994a).

The practical implications of this situation are not trivial. On the contrary, they arise every time we see a news story warning about the dangers of Alar on our apples, EDB in our breakfast cereal, artificial sweeteners in our soda pop, or asbestos in the buildings where we live and work. The reports typically include conflicting claims from environmentalists, manufacturers, and government officials. Unfortunately, journalists' views of who are the experts differ greatly from the views of the relevant expert community.

The activist ranks include some frequently quoted scientists whose views reflect the world of political activism better than that of scholarly research. Our survey of cancer researchers suggests that these individuals have more credibility with journalists than they do with their scientific peers. Ironically, however, it is their media-generated prominence that makes them appear more representative of the research community than they actually are.

Is the News a Health Hazard?

How well do the media communicate the dangers of cancer-causing agents in the environment? Do we get the information we need to minimize this widely feared health risk? The answers suggested by our results should give pause to anyone who follows news reports in order to assess the risks of cancer-causing agents in the environment.

Our central finding is that the factors stressed by the experts differ sharply from the dangers that we read or hear about most frequently. Our surveys of scientific experts largely echo the earlier Doll-Peto study of the field in identifying tobacco, diet, and sunlight as the primary contributors to the spread of cancer, while downplaying the role of artificial substances and industrial development.

By contrast, the national media have given heavy coverage to the cancer risks of man-made chemicals, food additives, pollution, radiation, pesticides, and hormone treatments. News stories targeted each of these factors as cancer agents more often than diet, sunlight, and asbestos. Indeed, food additives were named as carcinogens as often as diet and sunlight combined, and chemicals were cited more frequently than tobacco.

Scientific opinion stands in sharp contrast to the media-driven cancer scares that led the government to evacuate Love Canal and Times Beach after traces of dioxin were discovered there, forced grocers to pull breakfast cereals off their shelves to avoid EDB contamination, and panicked consumers into throwing out their applesauce for fear of Alar. In fact, the experts described all three of these notorious chemicals as causing fewer cancers than aflatoxin, a fungus found in peanut butter.

Moreover, the coverage failed to reflect expert opinion on several highly visible controversies about cancer causation. Six out of seven sources quoted in the media agreed that the United States faces a "cancer epidemic," an opinion shared by fewer than one in three scientists. Two out of three news sources argued that cancer-causing agents are unsafe at any dose, while two out of three scientists rejected this view. Nearly two out of three scientists rejected the government-mandated method of assessing human cancer risks by administering extremely high doses of suspected carcinogens to animals. By contrast, fully half of the media sources agreed that this is a good way to establish cancer risks for humans.

Thus, the national media have failed to convey expert assessments of environmental cancer risks to the general public. A regular consumer of news would get the impression that many man-made substances pose greater risks than most cancer researchers believe, while failing to appreciate the magnitude of risk that experts assign to such life-style factors as dietary choices and exposure to sunlight. The most careful reader or viewer would be misinformed about much of the scientific debate over cancer causation.

This is not to say that the experts are always right or that their opinions outweigh all other factors. But their views are an invaluable resource for policymakers and individual citizens concerned with minimizing health risks in a complex environment. When the media ignore expert views or report them inaccurately, the information environment can become hazardous to our health.

6
Things to Come

IF ONE IS TO BELIEVE THE RELEVANT scientific community, many environmental activists have grossly exaggerated the dangers of environmental cancer in their pronouncements about food, pesticides, pollution, clean water, and other topics. Studies of the research literature and surveys of scientists demonstrate this point. Cancer experts see much less danger from man-made substances such as pesticides than do environmental activists. They have relatively little respect for those scientists who speak for environmental organizations and they have little respect as well for the expertise of the organizations themselves. Indeed they rate groups such as the Environmental Defense Fund only slightly higher on a scale of reliability than the Tobacco Institute.[1]

They also give relatively low ratings to newspaper coverage of environmental issues, including the coverage offered by the *New York Times*. And no wonder. Newspapers tend to report the views of environmental activists as if they represented the views of the expert scientific community. So the scientists are correct in their assessment. The accuracy of newspaper coverage of the scientific issues involved in environmental cancer is, by any measure, quite poor.

Of course the scientific community may be wrong about environmental cancer, for a variety of reasons. Our data, however, do not support the argument that the views of scientists are corrupted by contacts with business or ideology, and the issues in this field are such that their work constitutes what Thomas Kuhn would have called normal science (Kuhn 1970), depending on replicable experiments and observations. While one cannot preclude the possibility, there is little evidence that some new paradigm is about to emerge which will dramatically reinterpret what the scientific community has uncovered.[2]

Recent history does offer examples of scientists being quite conservative in some areas of environmental concern and only gradually changing their opinions under the weight of evidence.[3] Certainly they are far from unanimous in their views on environmental cancer even now. Nevertheless, at a minimum, one would expect that journalists would describe expert views accurately, even if the scientists ultimately reject them. One can go further. We maintain that in choosing alternative regulatory policies (where costs are entailed), it is rational to rely upon the risk assessments of the scientific community, though not necessarily their policy recommendations.

Indeed our findings in themselves provide no clear regulatory standards. Scientists are not nearly as concerned about dioxin as are activists or some administrators in the EPA or the media. Yet a significant group of scientists do believe that dioxin poses serious risks. Further, the appropriate definition of acceptable risk is not clear, and though we have a few remarks to make about the issues below, the moral and practical problems are quite complex and beyond the scope of this study.

In any event, given the popular fears of cancer, the techniques upon which the environmental movement depends are very effective. Certainly that is part of the reason that the dangers of environmental cancer are stressed by organizations whose future depends, to some degree, upon uncovering environmental degradation and danger. But organizational self-interest is by no means the whole story. It is probably not even the most significant part, for the story reflects economic and social change as well as ideology.

The environmental movement is partly a response to real problems created by the success of the human species. Industrialization has greatly improved human health, as manifested by rapid population growth, but the by-products of this achievement have entailed considerable costs. The costs (and dangers) are likely to rise as less-developed nations such as China and India expand production, increasing the competition for relatively scarce resources and producing tremendous amounts of waste that contaminate the environment. These are issues with which the human community must deal. As we noted earlier, the greater sensitivity to the environment that activist organizations have engendered has had positive consequences in many areas. It is also clear that our very affluence and longer life spans have increased the relative importance of such late-developing diseases as cancer.

In addition, we are increasingly capable of uncovering smaller and smaller amounts of possibly dangerous materials in our environment. Many people, especially those not scientifically trained, regard any amount of a potentially dangerous substance as a threat to be dealt with. What's more, the public now widely adheres to the view that science can eliminate such materials at relatively little cost, so that we can afford a "pure" environment without sacrificing anything else. In short, as Robert J. Samuelson (1996) has pointed out, we continue to expect far more from the economy than it can realistically deliver.

All these factors have certainly influenced large numbers of environmental activists as well as the general population. Yet none of these, nor all of them combined, fully explains the beliefs and behaviors of the leading cadres of environmental activists. To understand them one must add something else.

As we already demonstrated in chapter 5, the contemporary environmental movement has partly been a surrogate for a general hostility to American capitalism, or at least suspicion of business and of businessmen and businesswomen. Thus the views of some environmental activists have derived to some extent from a broader set of political and social attitudes.

It should be noted, however, that while there are still some left-leaning radical groups, today's mainstream environmental leaders are not as far to the left as an earlier generation of mainstream environmental activists were. But they are well to the left of the general population.[4] Our survey reveals that they are relatively secular, suspicious of business, and overwhelmingly supportive of the liberal wing of the Democratic party. They favor extensive governmental economic intervention; they are also much more supportive of abortion rights, the rights of homosexuals, and preferential treatment for minority groups than is the population as a whole.[5] Thus, their belief that contemporary capitalism or contemporary capitalists produce and sell products that cause cancer is but one more manifestation of a generally critical view of their society. The discussion of environmental cancer, like other aspects of the environmental movement, has been partly politicized.

The cancer researchers in our sample (which is largely composed of academics and government scientists) also hold liberal views on social issues. Their political orientation does not, however, appear to carry over to their areas of expertise. Further, they are relatively moderate as compared to the environmental activists. Indeed the activists' views contrast sharply with those of the general population on a wide range of subjects, including environmental issues, despite the public's own continuing environmental concerns.[6]

Why have journalists at major media outlets turned to the environmental groups for news about the environment and believed them rather than others? There are clearly many reasons for such behavior, not the least of which is the ability of environmental groups to organize effectively and (though this can be exaggerated) journalists' penchant for "bad" news. The attraction to disaster, however, does not explain the media's willingness to side with the scientific establishment on the fluoridation controversy during the 1950s (Mazur 1981) or report the AIDS controversy with such care and sympathy, stressing that casual contact with an AIDS victim is not dangerous (Media Monitor 1987; Pope and Lichter 1989).[7]

As Mary Douglas and Aaron Wildavsky (1982) point out, the popular argument that the public's and journalists' concern about, say, pesticides is merely a function of a particular type of risk assessment engaged in by laymen as opposed to scientists is not persuasive. It is certainly true, as some have argued, that people are more willing to take risks they conceive of as being self-imposed (such as driving an automobile) than to accept those they perceive to be imposed on them by others (such as the use of pesticides). There is little evidence, however, that this affects *estimates* of risk, especially those of theoretically sophisticated journalists or of environmental activists who have been working in the field for some time.[8] Readers should remember that, in evaluating the dangers of cancer, the two groups differ only marginally on such substances as tobacco and most natural carcinogens (for example, fat in the diet). It is only when it comes to such man-made substances as pesticides and additives that the differences in danger estimates are considerable.

Some critics have argued that the problems of media reportage can be explained, paradoxically, by the very professionalism of journalists, who are trained to believe that there are two sides to every issue (Schudson 1995). But it is difficult to argue that professional norms lead to giving equal credence to positions that are rejected by most experts in the field, or to reporting on issues in a way that leads the public to believe falsely that most experts support a given risk assessment.

Others have argued that the problem is the scientific illiteracy of most journalists. Again we are not convinced. Journalists in the 1950s were less informed than they are today. The difference is that in the 1950s they gave much more weight to the views of the scientific establishment than they do today. This is a process which began in the 1960s and continues.[9]

We think that one of the most important reasons for journalists' poor reporting of environmental issues is their own beliefs, which are generally sympathetic to the views of environmental activists. Numerous studies (several of which are summarized in Lichter, Rothman, and Lichter 1986 and in Lichter 1996) have demonstrated the liberalism of elite journalists and their tendency to turn to public interest groups and "non-establishment" scientists for information on scientific issues (Goodfield 1981). We have also demonstrated that journalists are generally hostile to business, not only by comparing reporting on a variety of issues with journalists' ideology about the issues, but also by asking them to write headlines for stories or interpret various pictures we show them (Lichter, Rothman, and Lichter 1986). It seems reasonably clear that for journalists, political liberalism has gone hand in hand with support of environmental causes. In short, many journalists have taken their cues from environmentalists because they have found the fundamental ideological assumptions of such activists congenial. John Stossel, who reported ABC News's documentary "Are We Scaring Ourselves to Death?" explained his earlier credulity toward the claims of environmental groups: "we consumer reporters approached it from the bias that on the one hand is business, which is greedy and has an ulterior motive and will distort the data, and on the other hand is the noble environmental group, which has no other motive than to help the public" (Quoted by Shaw 1994c, part A, p. 32).

It is not that the best journalists do not attempt to balance their coverage. Yet many studies demonstrate that, when it comes to more ambiguous chains of events and over the long haul, their worldview (or ideological map)[10] affects what they see and report. This internal map is even more determinative for those journalists who, unaware of their own perspective, approach a story with the view that they are merely decent people who wish to comfort the afflicted and afflict the comfortable.

It has become difficult to deny any longer that, by and large, journalists are liberal (not radical) in their outlook. Still it is argued, among other things, that their professionalism keeps them from interpreting the world in a partisan manner (Dennis 1997).[11]

The Problem of Public Policy

The problem is not only that the news coverage of environmental cancer differs from expert opinion, leaving the public poorly informed about cancer risk. The deeper problem is that the public is being misinformed about the nature of expert opinion. The latter is a more serious impediment to an informed public as well as a direct indictment of the journalistic profession. We may not know how to achieve objectivity, but we can at least demand accuracy. While groups skeptical of some of the assertions of mainstream environmentalists are often characterized as conservative (for example, the American Council on Science and Health), environmental groups which are clearly liberal or even radical tend to be described simply as disinterested protectors of our environment (Smith 1994).

As we have argued, and a wide range of surveys demonstrates, many people have been persuaded by the media to believe that environmental problems are getting progressively worse and that America is facing a cancer epidemic. The Harvard Center for Risk Analysis (Graham 1995) found, for example, that the public gave "pesticide residues" a mean score of 6.9 on a 0-to-10 point scale. Compare this rating to the scientists' 4.72 rating on a similar scale.

These beliefs have had a number of undesirable consequences. The direct and indirect costs of regulating potentially dangerous substances are far greater than they have to be, taking resources away from those areas in which regulation is legitimate or from other public purposes. In 1992 James B. MacRea of the Office of Management and Budget concluded that OSHA was planning to spend an estimated $12.5 million to $20.4 million for the prevention of each premature death. He further concluded that more lives than that were probably being lost because reduced worker income was curtailing access to good nutrition, medical care, and safe products (Lave 1992). Pietro S. Nivola (1996) of the Brookings Governmental Studies program argues that regulation in the United States is far more detailed and adversarial than that of any other major industrial power. Such regulations and the litigiousness in tort cases permitted, if not encouraged, by the court decisions have powerfully and negatively affected productivity (Litan and Winston 1988). The American economy is carrying a heavy overhead that is a drag on living standards and our ability to support other worthwhile endeavors (Nivola 1996; *Economist* 1996b).[12] The Harvard Center for Risk Analysis (Harvard Group on Risk Management Reform 1995) estimates that regulatory costs on business (of which environmental costs are a major segment) may be as high as $600 billion a year. Of course, as they point out, there are also extensive benefits, amounting to perhaps $200 billion a year. The point is to maximize benefits relative to costs.

There is good evidence that the standards set for cleaning up waste dumps, insuring a clean water supply, or eliminating dangerous pesticides have been excessively high in terms of reduction of cancer risk (Margolis 1996; Easterbrook 1995; Schneider 1993a). For example, it is estimated that the rigidity of regulations on air pollution are costing as much as $15 million per potential life saved (Van Houtven and Cropper 1994). In pesticide regulation the implicit value of avoiding a cancer case is $52 million (Abelson 1993).[13] One of the most egregious recent examples of public hysteria leading to the major waste of resources was the removal of asbestos from buildings, the cost of which may have approached $100 billion. Such removal was mostly not necessary and may in fact have increased the risk of developing cancer, but the public had panicked (Abelson 1993). The same phenomenon partially explains the court-mandated awards by companies to Vietnam veterans for diseases, including cancer, supposedly produced by Agent Orange, despite the lack of evidence of any causal relationship (Gough 1991). The latest instance of this type has involved the large amount of money spent quite unnecessarily litigating the supposed dangers of breast implants on the basis of quite faulty data. While cancer was not the issue, the case is nonetheless instructive. Once again public panic led to large cash awards by sympathetic juries, awards that propelled a large company into bankruptcy (Kolata 1995a, 1995b; Angell 1996).[14]

It is certainly true that various industry groups have often understated the environmental dangers of their products and business operations. Over the years they have not been above presenting bad science to the public to advance their views and protect their profits. (Just think of the efforts of the tobacco industry.) Much of the public, however, regards business pronouncements skeptically, if not cynically. Thus, though their collective financial resources far exceed those of environmental groups, business's influence on public opinion is far smaller. Environmentalists may believe that they have accomplished little and are surrounded by powerful enemies, but any objective reading of contemporary history would have to emphasize the rapid pace at which the "greens" have come to dominate both public discourse and public policy on environmental issues (Easterbrook 1995; Chase 1995).

We are not, therefore, arguing that environmentalists are always wrong or that they are the only ones whose underlying social perspectives partly determine how they respond to evidence. It is clear that in a number of very important cases—for example, on dangers to the ozone layer or early clean-water initiatives—environmental activists have been right. It is also clear that even when they have been only partly right, they have often played a constructive public policy role. Further, they have made all of us (and we include ourselves) much more aware of the potential consequences of our behavior on the world in which we and our children must live. Given a balanced approach to environmental concerns, such awareness can only be of benefit to the human species. But as of now, such an approach characterizes neither public discourse nor public policy.

As John Graham and Jonathan Wiener (1995) have pointed out, resolving risk tradeoff is an extraordinarily difficult problem which requires accurate data as well as a broad view of the human environment, lest attempts to deal with one putative danger worsen our situation by having unexpected effects in other areas. What is not permissible is to ignore the need to examine tradeoffs with the best information we can muster.

What of the Future?

It is possible to modify public views on environmental regulation by proper information. The mass public's heightened awareness of risk is not a function of a different approach to risk assessment from that of scientists. Rather it reflects deep-seated fears created by those who dominate public opinion. John Doble's survey of four hundred people was designed to reflect a cross-section of the population before and after presenting them the views of scientific experts on various risks. He found substantial shifts in risk perception in the direction of those supported by the scientific community (Doble 1995).

Actually, there are modest signs of a new skepticism regarding some of the pronouncements of various environmental groups. In the past few years several journalists and scholars have been chastising the mainstream environmental movement in a manner to which its leaders are not accustomed. In a series of articles in the *New York Times* in 1993, Keith Schneider (1993a–c) essentially accused both environmental groups and the media of exaggerating the dangers posed by artificial chemicals to human beings and the physical environment. These articles were followed by David Shaw's (1994a–h) series of essays in the *Los Angeles Times* and a major television documentary, ABC's "Are We Scaring Ourselves to Death?" which made many of the same points. Similar arguments have been advanced in other articles and books by longtime supporters of environmental causes, including Gregg Easterbrook's *A Moment on the Earth* (1995) and Martin W. Lewis's *Green Delusions* (1992). Still other critiques, such as Charles T. Rubin's *The Green Crusade* (1994), have accused some members of the scientific community of becoming politicized, that is, of knowingly making judgments that do not comport with scientific evidence in order to support environmentalist initiatives.[15] Needless to say, many in the environmental movement have sharply challenged these revisionist arguments.

One does not know how much to credit the new journalism with public policy changes. Yet the new legislation emerging from Congress on pesticides, while not perfect, does set more reasonable standards. Instead of using the criteria mandated by the FDA's interpretation of the Delaney clause, pesticide manufactures will in the future be required to demonstrate only that that pesticide residues create no reasonable risk of harm (Cushman 1996b; *Congressional Quarterly* 1996).

Why have the attitudes of some journalists and political figures begun to shift? After all, the views of the scientific community were not very different in 1984, when we completed our first survey, from what they are today. Why have the national media, which were so unreceptive to science's message for so long, begun to show modest signs of change? That is another long story, which we can only discuss briefly here.

The environmental movement in its early stage was clearly even more ideological than it is now (Rothman and Lichter 1987). As it has become part of the establishment and as its composition has changed, it has become more subject to criticism. After all, as Easterbrook (1995) points out, the mainstream movement has won many of its battles, with results that can be easily seen. Beyond that, knowledge about cancer causation and its relation to public policy has been growing among journalists, partly stimulated by the environmental movement itself. The presence of an increasing number of scientifically trained environmentalists has also affected journalists, who have acquired more expertise themselves. For example, the highly regarded Center for Risk Analysis at the Harvard School of Public Health has become an important source of expert information for journalists, as have such environmentally moderate activist groups as the American Council for Science and Health.[16]

Finally, practicing scientists have begun to take a more active role in publicly assessing risks from various activities or substances. The 1995 report by the American Physical Society on the supposed relations between power lines and cancer is a good example of this development (*New York Times* 1995), as was the 1997 report (Kolata 1997a) on the relationship, or rather lack of relationship, between power lines and childhood leukemia.[17]

What can we conclude? It is clear that environmental groups were long able to capture the interest and support of the media for claims about the findings of scientists which were simply inaccurate. Whatever the reasons for this, the end result was that bad science outpointed good science, because ideology triumphed over evidence. Environmental cancer, as defined by environmental groups and the media, has been partly a political disease. And though a few scientists have lent their names to bad science, the great majority have not.

In some respects segments of the public are gradually taking a more realistic view of environmentally triggered cancer. While it would be nice to end this book on a positive note, it is not clear how lasting the recent improvement will be. For that matter, we are not even sure that rationality has won a victory. The attention paid to such revisionists as Easterbrook, Lewis, and Schneider does not mean that their views now dominate the mass media. Indeed, more journalists have attacked them than have supported them in such venues as journalism reviews and in many books. A not untypical example is Robert N. Proctor's *Cancer Wars* (1995), in which Bruce Ames and Samuel Epstein are treated as if they were equally respected in the field, and where the arguments made are largely ad hominem. It remains true, as well, that legislative efforts to develop more rational policy often founder as political figures beat a rapid retreat, lest they be attacked as hostile to the environment (Center for Risk Analysis 1995).

Moreover, as many have pointed out (e.g. Gross and Levitt 1994; Gross, Levitt, and Lewis 1996), a wave of anti-science is now sweeping segments of the academy, manifesting itself as "deep ecology," "ecofeminism," "social constructionism," and the like. While support for such ideologies is currently rather limited, these views appear to be gaining influence among a new generation of academics.

The growth of irrationality in the academy, as marginal as it still is to practical science, is disturbing. Just how serious are these trends for the future? Science does require a supportive social and political order. It can be crushed or inhibited by those in power, if they are sufficiently ideological and ruthless. Witness, for example, the destruction of genetics in the Soviet Union under Stalinist-supported Lysenkoism, as well as the longtime refusal of the Soviets to accept linear programming or even relativity theory for a variety of ideological reasons.

The growth of the anti-science movements, the very bad science successfully championed by various environmental groups, and the rise of various new-age groups espousing so-called spiritual, nonscientific ways of dealing with reality, including increased belief in astrology and magic, illustrate what a fragile endeavor science is for all its successes. Human beings do not naturally think scientifically. Indeed, as Jean Piaget and others have demonstrated, the thought of most children is at least partly magical. Rational methods of inquiry are learned only with difficulty (Wolpert 1993, Alcock 1995).

The point is well made by Phillip Gerrans (1997), who is worth quoting at some length:

> All cultures reproduce structures of violence, inequality and hierarchy rather than the distribution of power and resources according to any recognizable rational norm. Furthermore there is a lot of recent psychological research to show that this public irrationality reflects the structure of human unreason. Rationality is a precarious achievement, almost instantly swamped in day-to-day life by other influences on human psychology: selfishness, sexual desire, power and gratification in the short term. Furthermore, it seems that rationality, even in situations where it is recognized as desirable, such as planning and prediction, is not an automatic part of the human inferential repertoire. Humans are very bad at employing the norms of rationality such as logical inference, probabilistic reasoning and planning beyond the immediate future. . . . Even those arenas of human culture consecrated to the norms of rationality, the aca-

> demic professions, operate, in some cases, according to non principles of confabulation, improbability and counter-induction. (Gerrans 1997, pp. 4–5)[18]

It is not clear that the American public knows very much about even elementary scientific facts and concepts. A 1996 National Science Foundation study of a cross-section of Americans (*New York Times* 1996c) shows that only 9 percent of adult Americans know what a molecule is, and only 5 percent can give an explanation of acid rain. More startlingly, less than half know that the earth orbits the sun annually. It is little wonder that the struggle against the bending of science to serve ideal or material interests never ends.

Notes

Preface

1. For discussions of the history (often very partisan) of the "war on cancer," see, among other books, Strickland 1972, Patterson 1987, and Procter 1995. A brief discussion will be found in chapter 3 of this book. On cancer incidence and death rates, see Bailar and Gornick 1997. As this book was being edited, new and encouraging data on cancer incidence and morbidity was published in a joint report by the American Cancer Society, the National Cancer Institute (NCI), and the Center for Disease Control (CDC). Entitled "Progress Against Cancer: A Report to the Nation" (1998). Further studies have been published. On the decline of cancer rates see Wise, 1998. On progress in treatment of breast cancer see Altman, 1998; Harvard Health Letter, 1998; Nichols, 1998. On overall progress in seeking a cure for or seeking mechanisms for preventing cancer see Barinaga, 1998; Arap, Pasqualini and Ruoslahti, 1998; Scanlon and Kashani-Sabet, 1998.
2. Nevertheless, participants on both sides of the debate about environmental cancer accuse their opponents of rejecting the consensus of the scientific community and relying on "junk" science. See, for example, Ehrlich 1996 and Whelan 1992.

1: Historical Lessons of the Environmental Movement

1. For more detailed discussions of the impact of Marsh's work, see Botkin 1990, pp. 8–9; Petulla 1980; and Nash 1989, p. 38.
2. Many critics contend that Thoreau's version of wilderness is actually rather

civilized. During his famous retreat to Walden Pond, he lived only two miles from the town of Concord, where he ventured regularly. For one such critique, see Dubos 1980, pp. 14–16.

3. The two terms are actually a recent invention, used by Samuel Hays and other historians as a tool for tracing the historical development of the two ideologies originally contained within conservationism (Tucker 1982, pp. 54–55).
4. Critics argue that the original purposes of these reserves—human recreation, protecting water and game supplies—were anthropocentric and inconsistent with true preservationist goals (Nash 1989, p. 55).
5. The Hetch Hetchy controversy is discussed in almost every history of conservation. For more detailed accounts, see Nash 1982, pp. 161–181, and Fox 1981, pp. 139–147.
6. While the Nazi regime did put into practice some environmental ideas (such as organic farming), many of its policies, including re-armament, had a negative impact on the environment.
7. The products of the information age, however—personal computers, fax modems, etc.—are not intensive energy consumers.
8. For more detailed discussions of *Silent Spring,* Rachel Carson, and the impact of her book, see Rubin 1994, pp. 30–52; Nash 1989, pp. 78–82; Udall 1988, pp. 195–203.
9. Rejecting the term *pesticides* as making an anthropocentric value judgment that insects cause harm or annoyance to humans, Carson prefers the term *biocides,* to emphasize that such chemicals destroy life.
10. The effect of DDT in the environment is still controversial. While DDT was not an important carcinogen, it did adversely affect many species of birds and some other life forms. It seems, for example, to have been responsible for eggshell thinnings and consequent raptor and pelican declines in the 1950s and 1960s (Wildavsky 1995, pp. 55–80).
11. Ehrlich entered a debate with the economist Julian Simon, a critic of environmentalists, as to the near future. Ehrlich had predicted a general increase in the price of natural resources. Simon bet him that the prices for any natural resources chosen by Ehrlich would fall. Simon won the bet, but Ehrlich was unrepentant, dismissing his errors as unimportant for the long term. His defeat does not seem to have undermined his general credibility. For a discussion, see Rubin 1994.
12. The publication of *The Closing Circle* began an intense debate between Commoner and Ehrlich over the causes of ecological disaster. Beginning with an exchange published in *Environment* in 1972, Ehrlich maintained that overpopulation was the primary cause of the environmental problems facing the earth, whereas Commoner continued to emphasize the detrimental effects of technology and capitalism. For further discussion of the debate, see Paehlke 1989, pp. 58–62; McCormick 1989, pp. 69–73; Fox 1981, pp. 311–313.

13. For more detailed discussions of this study, see Paehlke 1989, pp. 50–54; McCormick 1989, pp. 74–79; and Rubin 1994, pp. 130–141.
14. Perhaps the most famous critique of *Limits* is Cole et al. 1973, published specifically in response to the Club of Rome study. Other detailed critiques include Tucker 1982, pp. 194–211 and 215–219, who discusses the technical problems with the study, and Simon 1981, who argues against the hypothesis of limited resources.
15. Popular accounts of environmentalism generally name Earth Day as the beginning of the environmental movement.
16. The policy committee names—Water Resources, Biodiversity, Political Education, Energy, Economics, Environment and Education, Air Quality—demonstrate the breadth of issues now addressed by the Sierra Club.
17. Critics contend that the risks posed by the Hooker Chemical Company waste dump were grossly exaggerated. Although the government spent $50 million on evacuation and clean-up, a 1991 article asserted that "to date, little or no scientific evidence has been produced to justify the Love Canal panic" (Klaidman 1991, p. 77). Since the disaster, residents have resettled the evacuated area. For a more recent analysis, which cites further evidence of the exaggerations of the media, see Easterbrook 1995. As Easterbrook also points out, many of the other so-called disasters that received major press coverage were not disasters at all. No one died at Three Mile Island, and the oil spills were not nearly as catastrophic as the media implied. Even the damage from the Exxon Valdez accident in Alaska was quite minimal, despite the fact it is regarded as a sacred text by some environmental groups. As Easterbrook points out, however, the area contaminated by the spill had recovered as of 1995, and the nontreated segment of the beaches have recovered more fully than those that were treated. Exxon spent some $2 billion cleaning the site in a vain attempt to win a more favorable public image. Cases like this have led to a reaction against the environmental movement on the part of some journalists, a reaction we discuss in the concluding chapter.

2: Understanding Contemporary Environmentalism

1. For a detailed description of ecocentrism and technocentrism (or resourcism) see Oelschlaeger 1991 and O'Riordan 1981. We are aware that different systems of classification are in use. Thus Easterbrook 1995 distinguishes ecorealism from ecological pessimism, and Lewis 1992 distinguishes between mainstream and radical environmentalism. The distinctions are all attempts to make the data with which we deal somewhat more accessible. None are completely satisfactory.
2. The term *ideology* is not used to disparage the movement. We follow Clifford Geertz's definition of ideology as a map of empirical and moral rules designed to guide us when we attempt to act in the world. Thus all of us rely upon an ideology (Geertz 1973). Synonyms include such terms as *perspective,*

paradigm, and *outlook,* though the first and third terms do not really suggest as integrated a process as does *ideology.*

3. Nelson 1996 traces these attitudes, in America at least, to a secularized Puritanism.
4. This is something of an oversimplification. Deep ecologists and ecofeminists may be on the same side of the continuum. They differ, however, on many issues. For a summary that attempts to bridge some of the gap see Zimmerman 1994.
5. Snow 1992, pp. 15–21, identifies eleven broad categories: (1) small, all-volunteer issues groups, (2) small, quasi-volunteer naturalist groups, (3) recreation and sporting clubs, (4) state or regional advocacy groups, (5) education, research, and policy development centers, (6) law and science groups, (7/8) small and large national and international membership groups, (9) real estate conservation groups, (10) professional societies, and (11) support and service organizations.
6. We calculated this number by adding together all of the associations listed under the following subject headings: conservation, ecology, environment, public lands, range lands, waste, wildlife conservation, animal welfare, and parks and recreation.
7. For a more detailed examination of the campaign for the sea-dumping ban, see Specter 1993. Rosen 1990 examines the tension between science and public relations demonstrated in the Alar scare. For revisionist views of the hazards of asbestos, see Stevens 1990a, and Hooper 1990. Discussions of the politics of the landfill debate include Rathje and Murphy 1992 and Passell 1991a.
8. Thomas DiLorenzo 1990 summarizes the socialism-pollution link in his rebuttal to the arguments linking capitalism and pollution. For a detailed examination of Commoner's critique of capitalism, see Rubin 1989, 1994. Stretton's 1976 book arguing for the superiority of socialism is not untypical. A survey by Lichter, Lichter, and Rothman 1983 found that a majority of public interest leaders thought it would be a good thing if the United States moved toward socialism. They also were more supportive of Fidel Castro than of Ronald Reagan.
9. Recent assessments of environmental destruction in the former Communist countries are devastating. The former deputy chairman of the Committee of Environmental Problems in the Soviet Union admits that about 45 million Russians live in ecological disaster zones and that environmental degradation accounts for much of the decreased life expectancy of Russian men (Yablokov 1993, p. 579).
10. Other recent works examining environmental destruction in the former Soviet Union and its satellites include the October 1993 special issue of *Environmental Science and Technology,* the September 1990 issue of *Options,* and Ziegler 1987, 1991.
11. To be fair, many environmentalists with socialist leanings did not regard the Soviet Union as a model for anything. They claimed that a more decen-

tralized, more democratic socialism could be created which would be more sound environmentally than capitalism is. See Lewis 1992, pp. 150–190.

12. Early works critical of nuclear power include Novick 1969 and Curtis and Hogan 1969.
13. On this and the following paragraph, see the following: Easterbrook 1995, pp. 492–526; Malhin 1995, Cohen 1990, Colglazier 1982, and Westerstahl and Johansson 1991.
14. All of the senators and all but one representative voted for the Clean Air Act of 1970 (Tobin 1984, p. 227).
15. For much of the previous section we have relied on Kenski and Ingram 1986, who provide a useful chart tracing the evolution of pollution policy.
16. For more detail about the Clean Water Act and related legislation, see Ingram and Mann 1984, pp. 253–263.
17. From early on at least some commentators argued that although regulation reduced automobile emissions, factory-produced pollution had been declining all along and it was not clear that government regulation achieved much (Crandall 1981).
18. For the academic debate, see Pope et al. 1995, Cohen and Pope 1995, Stevens and Moolgavkar 1984, Samet and Cohen 1995, Moolgavkar et al. 1995, Moolgavkar and Luebeck 1996, Samet 1996, Tango 1994, Ames and Gold 1997, and Ames, Gold, and Willett 1995. For more popular essays, see *The Economist,* March 15, 1997, Bilirakis 1997, Baucus 1997, Milloy and Gough 1997, and Fumento 1997. Michael Fumento's book is rather polemical, but he makes some excellent points and provides additional references. A further discussion will be found in chapter 3, p. 78. As we note in chapter 4, the vast majority of experts in the field still do not consider air pollution a significant source of cancer.
19. For evaluations by scientists who stand at opposite poles on this issue, see Ehrlich and Ehrlich 1996 and Ames, Gold, and Willett 1995 and Ames and Gold 1997.
20. The maximum tolerable dosage is that just below the dosage which would cause an animal to die because of the toxicity. The Delaney clause, passed in 1958, absolutely bans any food additives that have been found to "induce cancer in man or in animals." See chapter 3 for further discussion of the Delaney clause and the consequences for public policy.
21. For a full discussion of these issues and general concerns about forest land and timber cutting, see Chase 1995.
22. CFCs are considered one of the major causes of ozone depletion.
23. For the above discussion, see Wildavsky 1995, Easterbrook 1995, Firor 1990, Fisher 1990, Schneider 1988, Stern, Young, and Druckman 1992, and Cline 1992.
24. For further discussion see Mahoney 1990 and Wildavsky 1995.
25. Of course, "deep ecologists" loath the notion of sustainable development. See also Lewis 1992.
26. Articles praising and critiquing ecofeminism appear regularly in *Environ-*

mental Ethics. For detailed discussions of ecofeminism and related topics see the Winter 1991 special edition of *Hypatia,* Warren 1990, and Warren and Cheney 1991. For a sharp critique, see Gross and Levitt 1994.

27. Other ecofeminists conclude that women's better understanding of nature is a product of a culture which has constructed their lives that way.
28. Differing perspectives on radical environmentalism are found in Manes 1990, Lewis 1992, and Rubin 1994. Of course, Gross and Levitt 1994 are quite hostile.
29. There is no evidence that traditional or primitive peoples, including Amerindians, were more concerned with the environment than European Christians have been, despite the belief that even inanimate objects possessed spirits. The mythology created by radical environmentalists, however, does not necessarily undercut their ultimate goals. For a discussion of these matters see Low 1996.
30. Of course if too great an imbalance is produced, the result may well be the overall collapse of the system.
31. See the discussion in chapter 6, pp. 181–185.

3: What Is Environmental Cancer?

1. On misperception of risk, see Fischoff 1985, Slovic 1987, and Wildavsky and Dake 1990. More generally, see Proctor 1995.
2. These remarks were made by Sen. John Tunney, speaking at an October 24, 1975, Senate Commerce Committee (Subcommittee on Environment) hearing on the Toxic Substances Control Act. Tunney was reflecting a legitimate estimate that had been distorted for shock effect; see the discussion of an estimate of "environmental cancer" below, p. 00.
3. In our review, we have relied upon Doll and Peto 1981; Office of Technology Assessment 1981, a congressional report for which Doll and Peto were the lead consultants; the work of Bruce Ames, a biochemist at the University of California, Berkeley, particularly Ames 1983, Ames, Magaw, and Gold 1987, Ames, Gold, and Willett 1995, and Ames and Gold 1997. For reviews of recent cancer statistics see Devesa et al. 1995 and Bailor and Gornick 1997. A quite different view is presented by Proctor 1995.

   The scholars we relied upon were chosen partly because our survey found that they were held in the highest regard by cancer experts. As the reader will discover in chapters 4 and 5, their summaries do seem, by and large, to reflect the views of the expert community. We must point out, however, that a number of reputable scientists take a much more cautious view about, say, pesticides than does Ames: they include John Wargo (see, for example, Wargo 1996 and the sources cited therein), David Rall, Michael Gallo, and Bernard Goldstein (see Lave and Upton 1987).
4. This early work is cited by David S. Fischer in "The Etiologies of Cancer," in Fischer 1982, p. 12. This volume is one of the standard medical school texts on cancer.

5. This history is summarized from Richards 1972, sections 1–2, and Glasser 1976, pp. 8–10.
6. On the Delaney amendment, see Salsburg and Heath 1981, pp. 32–33.
7. See Ames 1984, remarks on *The McNeil-Lehrer News Hour,* pp. 9–10; Ames et al. 1987; and the profile of Ames in Proctor 1995, chap. 6.
8. For more on national differences in cancer rates, see Newell et al. 1982; Hellman and Rosenberg 1982, pp. 18–21; Doll and Peto 1981, pp. 1198–1200; and more generally Doll et al. 1966.
9. On the influence of migration on cancer types and rates among Japanese-Americans, see Newell et al. 1982, pp. 19–21; on African-Americans, see Doll and Peto 1981, pp. 1200–1201.
10. For sources, see n. 3, above.
11. Though later researchers have modified some of their analyses and estimates, most of what they wrote stands up today, and the work is regarded as classic.
12. In its 1987 report *Unfinished Business,* the EPA made new estimates of the number of cancer cases associated with exposures from various sources. Gough 1989 found these to be remarkably similar to the Doll and Peto estimates. He later argued (Gough 1990) that, based on the EPA's own data, its existing programs, even if perfectly administered, would eliminate very few cases of cancer. For an update, see Devesa et al. 1995, which found increases from 1975 to 1991 in prostate and skin cancer among men, and breast and lung cancer among women. The increases in skin cancer were often associated with immune deficiencies brought on by AIDS.
13. See Higginson and Muir 1979. These patterns, with some variation, were found again by Devesa et al. 1995.
14. On the relation of diet and cancer, see Berg 1976; Doll and Peto 1981, pp. 1226–1235; Office of Technology Assessment 1981, pp. 76–84; Schapira 1992; Statland 1992; Whelan 1994, pp. 284–286; National Research Council 1993, 1996.
15. Ames and Gold (1997) have come to the conclusion that a high intake of fruits and vegetables contributes significantly to cancer prevention.
16. The politics of tobacco are discussed by Frischler 1989 and Klugar 1996.
17. At least one court ruled that because cigarettes contained a drug they could be regulated by the FDA (Weiser 1996).
18. See, for example, Klugar 1996 and Hilts 1996.
19. See Rothman and Keller 1972; cited in Doll and Peto 1981, pp. 1224–1225.
20. The scientists listed were all in prominent positions within government agencies that have responsibility in the area of occupational cancer; Kenneth Bridford, NIOSH; Pierre Decoulfle, NCI; Joseph F. Fraumeni, Jr., NIOSH; David Hoel, NIEHS; Robert N. Hoover, NCI; David P. Rall, NIEHS (director); Umberto Saffiotti, NCI; Marvin A. Schneiderman, NCI; Arthur C. Upton, NCI (director).

21. On policy changes of this type that cannot be explained by reference to the material goals of those involved, public choice theory notwithstanding, see Derthick and Quirk 1985.
22. Note that between 1975 and 1991, deaths from non-melanoma skin cancers increased by more than 800 percent, but the rapid increase is explained by increases in Kaposi's sarcoma, an opportunistic cancer common among AIDS patients; see Devesa et al. 1995, pp. 175–176. For baseline figures, see Doll and Peto 1981, pp. 1253–1254.
23. This overview of breast cancer risk was drawn from Doll and Peto 1981, pp. 1237–1238; Office of Technology Assessment 1981, pp. 99–102; Whelan 1994, pp. 192–95; Peto et al. 1996; Eeles et al. 1994; Borresen 1992; and *Tufts University Diet and Nutrition Letter* 1996b.
24. On prostate cancer, see Newell et al. 1982, p. 21, and the sources on reproductive system cancers cited above.
25. A study of the fifty largest newspapers in the United States by the National Cancer Institute found that the most frequently mentioned causes of cancer from 1977 to 1980 included food additives, pollution, and chemicals. See Boffey 1984.
26. In this regard I am following Harris, Nicholas, and Milvy 1981, p. 17, and National Research Council 1996, as well as Houben et al. 1992 and Lin 1992.
27. There is a large literature on the regulation of artificial sweeteners; see Havender 1983b; Doll and Peto 1981, p. 1236; Office of Technology Assessment 1981, p. 83; Wildavsky 1995, pp.26–37; Mahoney 1992; National Research Council 1996.
28. The contrast between regulation of saccharin in the United States and Canada is instructive on this point; see Harrison and Hoberg 1994, chap. 5, "Paternalism vs. Consumer Choice."
29. Air pollution is not nearly as damaging as smoking in part because smokers draw carcinogens from cigarettes directly into their lungs. However, on a bad day at the LaBrea tar pits, visitors breath through their noses which filter out most impurities. For references on issues of air pollution, see chap. 2, n. 18. Some recent studies suggest a stronger contribution to death rates by air pollution than had hitherto been suspected, but this is not certain. For discussions of environmental tobacco smoke, see Morris 1995, Miller et al. 1994, Tredaniel et al. 1994, Fontham et al. 1993, and Siegel 1993.
30. See Doll and Peto 1981, pp. 1249–1250; on the continuing controversy over chlorination, see Brown 1995.
31. One of the problems with using DDT or other pesticides is that some insect species mutate and develop immunity.
32. See Christine Russell 1990; for a comparison of the American and Canadian responses to these alarms, see Harrison and Hoberg 1994, chap. 4.
33. See *Columbia Journalism Review,* September/October 1996, pp. 13–15; November/December 1996, p.7. Our own survey of cancer experts supports Whelan's contention.

34. But see Wargo 1996.
35. For details of the litigation, see Schuck 1986.
36. "Crude rates" are calculated without taking into account age changes in the population. Use of crude rates gives the appearance of higher cancer rates. See Office of Technology Assessment 1981, p. 36.
37. The First and Second National Cancer Surveys were at one time known as the Ten Cities Surveys because the data base was from incidence statistics from Atlanta, Birmingham, Chicago, Dallas–Fort Worth, Denver, Detroit, New Orleans, Philadelphia, Pittsburgh, and San Francisco–Oakland. The Third National Cancer Survey added the Minneapolis–Saint Paul area and the states of Colorado and Iowa. The SEER program gathers data from Atlanta, Detroit, New Orleans, San Francisco–Oakland, Seattle-Tacoma, as well as Connecticut, Hawaii, Iowa, New Mexico, Puerto Rico, and Utah. The SEER sites were selected in part to address the underreporting of cancer among minority groups in earlier surveys. See Office of Technology Assessment 1981, pp. 37–40.
38. See Devesa et al. 1995 and the section on sexual behavior and cancer above.
39. Higginson's estimate is discussed by Proctor 1995, p. 56.
40. See the March 12, 1998, "Progress Against Cancer: A Report to the Nation" published jointly by the American Cancer Society, the NCI, and the CDC, which shows that cancer rates are declining for all sites and across most of the population as are the death rates for most types of cancer.
41. This is the essence of the position taken by Bruce Ames; see particularly Ames, Magaw, and Gold 1987; compare Proctor 1995, chap. 6, for criticism of that position.
42. The "issue network" model was introduced by Heclo in 1978.
43. For the traditional statement, see Freeman 1965 and Rourke 1984.
44. See Price 1962, p. 71, and Price 1965.
45. It is, in one sense, the network that Proctor 1995 describes in *Cancer Wars*.
46. Ackerman and Hassler 1981 provide an excellent case study of the use and abuse of these technology-based standards for controlling risk in *Clean Coal / Dirty Air*.
47. Ottoboni is a toxicologist with the California Department of Health Services.
48. In 1996 the Commission on Risk Assessment and Risk Analysis, which was appointed by Congress, recommended a number of changes in the manner in which risks are measured, including the repeal of the Delaney Clause (Kolata 1996c).
49. For an extended example in which this was done, see Melnick 1983.
50. See the profile of Waxman in Barone and Ujifusa 1991, pp. 140–143, and 1995, pp. 166–170.
51. See the profile of Bilirakis in Barone and Ujifusa 1995, pp. 315–317.
52. On Bliley, see Barone and Ujifusa 1995, pp. 1384–1386.
53. Enacting such limits is a part of the "Contract with America" promoted by

the House leadership in the 1994 election and the first few months of the subsequent congressional session; see, for example, Cushman 1995.

54. See, for example, Berke 1990. Current polling suggests a slight decline in support for environmental protection, but only that; for example, on April 7, 1995, the *Portland Oregonian* reported that both the *New York Times* and *Washington Post* polls showed popular opposition to most specific Republican policy proposals; see *Oregonian* 1995.
55. See Russell 1990, p. 7, on findings in surveys by industry. Indeed, so tainted are industry views that environmental activists often make their case merely by asserting that organizations which disagree with them, such as the American Council of Science and Health, are funded by polluters. See, for example, Ehrlich 1996, p. 161, and *Columbia Journalism Review,* September/October 1996, pp. 13–15; November/December 1996, p. 7.
56. Quoted from interviews by Steven Kelman 1981, p. 113.

4: The Experts versus the Activists

1. The shift goes back to the 1960s. See Goodfield 1981.
2. Surveys of opinion have many limitations. They are likely to be least accurate or predictive when individuals are asked what they would do under particular circumstances, i.e. fight for one's country "right or wrong," sacrifice one's self-interest in the interest of the environment, or spend time as a volunteer for a particular cause. As we have indicated, however, even straightforward questions may not be as clear as one hopes despite the fact that one has pre-tested them on comparable respondents. Our faith in our results derives from the fact that, as our readers shall see, the responses to a whole variety of questions point in the same direction. It is unlikely that all our questions shared an ambiguity that pushed the results in one direction.
3. As always, after the fact, we realize that the questions pertaining to secondhand tobacco smoke and asbestos may have been ambiguous. We let the scientists define what risk means in evaluating various substances. Thus, a review of the literature indicates that most scientists believe that asbestos was a major cause of cancer for those who worked with it in shipyards and other workplaces but not for the general public; and, though opinion is divided, secondhand tobacco smoke is believed to be dangerous primarily for those heavily exposed to it over time. We suspect that, to at least some extent, the responses of our sample as regards these substances reflect a perspective that defines danger in terms of extended exposure rather than a contribution to the overall cancer rate. While we cannot be sure, the responses to most of the other questions suggest that a "conservative" interpretation of scientists' views on these two issues is more likely to be correct than is a broader interpretation.

   We made still another error. We should have asked for field of specialization. As in so many other areas, research in environmental cancer produces its own specializations. One suspects that experts on carcinogens who have not themselves studied, for example, DDT are somewhat more likely to derive their views on its dangers from the media than are scientists who specialize in the study of pesticides.

4. The estimate is conservative because researchers were asked to list the number of articles published, the top cutoff point being fifty or more. Since the median score was fifty we can assume that a substantial number of scientists had published a larger number of articles than that. Given the reasonably strong negative correlation between publication and high estimates of risk, there is good reason to believe that the high and low ends of the scales would have been differentiated by significantly more than one point if the question had allowed us to adequately tap the attitudes of the most productive scholars.
5. The relatively small size of the sample also played a role. The smaller the sample, the larger the score difference has to be to achieve statistical significance. It should be noted, however, that even in general population samples women score as more risk-aversive than men, again by about a point on an eight-point scale (Harvard Center for Risk Analysis 1996b).
6. Here is a case in which a minority of scientists, described by the media as if they were a majority, seem to have been more correct in their estimates than their more skeptical colleagues. For further discussion, see chap. 2, pp. 44–45.
7. It is only fair to note that many environmental activists think of Easterbrook as an apostate rather than as a friendly critic. He has been attacked quite sharply since the publication of his 1995 book.

5: Media Coverage of Environmental Cancer

1. Coverage of tobacco sharply escalated in the spring and summer of 1997. As noted earlier, however, we lack systematic data on this period.
2. As pointed out in chapter 3, p. 81, at least some journalists still deny that the Alar scare was exaggerated, attacking those who so argue as the lackeys of big business.
3. Actually the NRDC is still highly regarded by the media. Its recent press release on the dangers of small particulates as a measure of danger from air pollution was respectfully received and provided the basis for an editorial in the *New York Times* (1996a, 1996b), though the data was based on a soon-to-be-released EPA summary of the research literature.
4. Both remarks are quoted in Brooks (1989).

6: Things to Come

1. It is rather ironic that activists such as Ehrlich have attacked those who criticize some of the contentions of environmentalists as the ones who are engaged in the "betrayal of science and reason," to quote the title of his 1996 book. We can not judge with certainty his contentions in other areas, but his claim that his chapter on pesticides reflects the views of the scien-

tific community is clearly false. The same may be said of the 1996 report of the Democratic caucus of the Committee on Science in the House of Representatives.

2. One might suspect that scientists today would be pushed in the direction of finding environmental cancer triggers. After all, that is probably where most of the government money (and most cancer research by university scientists depends upon government grants) is today. Our data, however, do not permit us to speak to this issue.
3. As we point out in chapters 2 and 4, global warming may be one such issue. Support for the hypothesis that the earth is warming and that human activity plays some role in that warming has slowly increased in the relevant scientific community as new evidence has been amassed. The issue is still highly controversial, however. In addition, as we have seen, new evidence seems to be emerging which suggests that air pollution is more dangerous toward one's health than had hitherto been believed. But it is too early to be sure.
4. For a discussion of the earlier generation of activists see Rothman 1994 and Lewis 1992. For discussions of the contemporary environmentalist fringe see Lewis 1992 and Gross and Levitt 1994.
5. We are not judging the rightness of these beliefs, some of which we share. We are suggesting that they are human constructions and not views that all right-thinking people hold because they are part of the nature of things. We note them despite our agreement with those who urge that government has an important role to play in ensuring that environmental concerns are addressed. At issue for us are only the extent of such regulation and the manner in which it is applied.
6. For a recent survey see Ladd and Bowman 1995. In addition to these findings, Rothman and other colleagues surveyed various leadership groups in American society. Among these groups we find a strong correlation between liberalism and belief in the seriousness of environmental problems (Lerner, Nagai, and Rothman 1996; Rothman and Black forthcoming).
7. The fluoridation took place in an era in which journalists deferred to the scientific establishment and its relatively small number of spokesmen. (See the discussion and reference below.) There were no activist environmental groups to act as intermediaries between the public and that establishment. Those opposing fluoridation tended to come from small towns, especially in the South, and were apt to talk about Communist plots designed to destroy America. In the minds of journalists who, by and large, still believed in scientific progress as an unmitigated good, such people seemed quite retrograde. In the case of AIDS, it is clear that journalists identified with homosexuals with regard to gay rights issues and viewed them as victims.
8. For a short review of the psychological literature on the differences in outlook regarding risk assessment between laymen and experts see Gillette and Krier 1990.

9. Journalists' turn away from the scientific establishment and toward various activist groups is described by Goodfield 1981.
10. For a view of the nature and role of ideology which we share see Geertz 1973.
11. Dennis's arguments are unpersuasive. He dismisses evidence rather than confronting it, and he is not untypical in this regard. Interestingly enough, public opinion surveys reveal that even media consumers have come to acknowledge the tilt of the press. More self-identified liberals in the general public perceive a liberal tilt in news reporting than perceive a conservative tilt. *Media Monitor,* May/June 1997. While this book was being copyedited, the First Amendment Center published a volume entitled *Worlds Apart* (Hartz and Chappell 1997), which documents the relative illiteracy of both the public and journalists on scientific issues and scientists' relatively negative view of journalists' coverage of science. As already noted, we do not deny that many journalists are scientifically illiterate. However, they were probably even more illiterate in the 1950s. The difference is that, then, there was a scientific establishment to which they deferred. Today, at the intersection between science and public policy, journalists tend to defer to various groups with axes like theirs to grind. How else does one explain the continued conviction on the part of a large number of journalists that science has established that silicone breast implants cause connective-tissue disease?
12. We have already discussed Superfund, but see Viscusi and Hamilton 1996. Evidence also supports the notion that recycling, the current feel-good activity of American culture, is also an expensive waste of time and effort (Tierney 1996). For an alternate view on risk assessment by an OSHA official see Finkel 1996a and 1996b. His assessments are out of line with those of most experts as revealed in our survey.
13. Porter and van der Linde 1995 argue that because of savings, pollution abatement in the United States is not very expensive. K. Palmer, Oates, and Portney 1995 are skeptical. They estimate total costs per year at about $100 billion.
14. Breast cancer has been particularly prone to relatively expensive and time-consuming study after study. In late 1997 a large-scale study found no evidence that exposure to either PCBs or DDT increased the risk of breast cancer. Naturally, various environmental groups raised the possibility that the study was unsatisfactory and called for still more studies. As one scientist responded, "For advocates, it's never-ending. But for other people there may be times when we want to spend our money on other things" (Kolata 1997b).
15. While this may be true in a few cases, our evidence does not, on the whole, support Rubin's view.
16. Of course, the Council is not consulted by many journalists who believe it is environmentally conservative.
17. The *New York Times* also provides a good example of the fragmentation of

journalism on issues of risk. In 1996 Dr. Theo Colborn, Dianne Dumanoski, and Dr. John Peterson Myers published a book with E. C. Dutton entitled *Our Stolen Future*. The laudatory preface was written by Vice President Al Gore, and the authors maintained, among other things, that various synthetic chemicals were disrupting male sperm counts. The book was favorably reviewed in the *New York Times Book Review* by a nonscientist who sympathized with the book and praised it highly (Hertsgaard 1996). Alerted by scientists, the *New York Times* reporter Gina Kolata found that a substantial number of prominent researchers had significant reservations about the book (Kolata 1996a). In addition, recent studies indicated that the supposed fall in sperm counts (an important source of evidence for the argument of the book) was not in fact occurring or, at least, the assertion that it was lacked credible evidence (Kolata 1996b; Rhomberg and Hernandez 1997; *Science News* 1996; *Vital Statistics* 1997). At a minimum the authors wrote with an assurance that was not warranted. For further discussion see Murray 1996b.

18. This is one good reason for studies such as ours. Though they often do not make for a very good read, they provide for the possibility of replication and public checking.

Bibliography

Abelson, Philip H. 1993. "The High Cost of Exaggerating Health Risks." *Public Perspective,* July/August: 3–7.

Abramson, Rudy. 1992. "U.S. Speeds Plan to Avert Ozone Harm." *Los Angeles Times,* February 12, p. A1.

Ackerman, Bruce A., and William T. Hassler. 1981. *Clean Coal/Dirty Air.* New Haven: Yale University Press.

Agran, Larry. 1977. *The Cancer Connection.* Boston: Houghton-Mifflin.

Alcock, James E. 1995. "The Belief Engine." *Skeptical Inquirer,* May/June: 14–18.

Altman, Lawrence K. 1998. "Studies Show Another Drug Can Prevent Breast Cancer." *New York Times,* April 21, p. A16.

American Cancer Society. 1998. "Progress Against Cancer: A Report to the Nation."

American Enterprise. 1995. "Environmental Protection, More Environmental Protection." March/April: 108–109.

Ames, Bruce N. 1983. "Dietary Carcinogens and Anticarcinogens." *Science* 221 (September 23): 1256–1263.

——. 1984. Remarks on *The McNeil-Lehrer News Hour,* March 2. Transcript from WNET, New York.

——. 1991. "Of Mice and Men: Finding Cancer's Causes." *Reason,* December: 18–22.

——. 1992a. "Science and the Environment." Unpublished manuscript. University of California, Berkeley, Department of Biochemistry.

——. 1992b. "Pollution, Pesticides, and Cancer." *Journal of AOAC International*

75 (1): 1–5. Keynote address at the 105th AOAC International meeting, August 12–15, 1991, Phoenix, Ariz.

Ames, Bruce N., and Lois Swirsky Gold. 1990. "Chemical Carcinogenesis: Too Many Rodent Carcinogens." *Proceedings of the National Academy of Sciences USA* 87 (October): 7772–7776.

——. 1997. "Pollution Pesticides and Cancer: Misconceptions." *Testimony U.S. Senate, Environmental Risk Factors for Cancer.* March 6.

Ames, Bruce N., Lois Swirsky Gold, and Walter C. Willett. 1995. "The Causes and Prevention of Cancer." *Proceedings of the National Academy of Sciences USA* 92 (June 1995): 5258–5265.

Ames, Bruce N., Renae Magaw, and Lois Swirsky Gold. 1987. "Ranking Possible Carcinogenic Hazards." *Science* 236 (April 17): 271–280.

Ames, Bruce N., Margie Profet, and Lois Swirsky Gold. 1990a. "Nature's Chemicals and Synthetic Chemicals: Comparative Toxicology." *Proceedings of the National Academy of Sciences USA* 87 (October): 7777–7781.

——. 1990b. "Dietary Pesticides (99.99% All Natural)." *Proceedings of the National Academy of Sciences USA* 87 (October): 7782–7786.

Angell, Marcia. 1996. *Science on Trial: The Clash of Medical Evidence and the Law in the Breast Implant Case.* New York: Norton.

Arap, Wadih, Renata Pasualini, and Erkki Ruoslahti. 1998. "Cancer Treatment by Targeted Drug Delivery to Tumor Vasculature in a Mouse Model." *Science,* 279, (January 16): 377–380.

Bailar, John C., and Heather Gornik. 1997. "Cancer Undefeated." *New England Journal of Medicine* 33, no. 22 (May 29): 1569–1574.

Bailar III, John C., and H. L. Gornik. 1997. "Trends in Cancer Mortality: Perspectives from Italy and the United States." *La Medicina del Lavoro,* 88, no. 4: 274.

Barinaga, Marcia. 1998. "Peptide-guided Cancer Drugs Show Promise in Mice," *Science,* Vol. 279, (January 16): 323–324.

Barone, Michael, and Grant Ujifusa. 1991. *The Almanac of American Politics, 1992.* Washington, D.C.: National Journal.

——. 1995. *The Almanac of American Politics, 1996.* Washington, D.C.: National Journal.

Baucus, Max. 1997. "Scientists See Need for New Air Standards but Disagree on Levels; So EPA Has Decided." Advertising supplement to the *Washington Post National Weekly Edition,* April 14, p. S2.

Berg, John W. 1976. "Nutrition and Cancer." *Seminars in Oncology,* March: 17–23.

Berke, Richard L. 1997. "In a Reversal, G.O.P. Courts the 'Greens.'" *New York Times,* July 2, pp. A1, A20.

Berke, Richard L. 1990. "Oratory of Environmentalism Becomes the Sound of Politics." *New York Times,* April 17: 1ff.

Bilirakis, Michael. 1997. "Neither Science nor Law Requires These Standards."

Advertising supplement to the *Washington Post National Weekly Edition,* April 14, p. S2.

Blumenthal, Herman 1978. "The Cancer Lottery." *Harpers,* Sept.: 12–21.

Boffey, Phillip M. 1984. "After Years of Cancer Alarms, Progress amid the Mistakes." *New York Times,* March 20: 17.

Bonner, Raymond. 1993. "Crying Wolf over Elephants: How the International Wildlife Community Got Stampeded into Banning Ivory." *New York Times Magazine,* February 7: 16ff.

Borrelli, Peter. 1988. *Crossroads: Environmental Priorities for the Future.* Washington, D.C.: Island Press.

Borresen, A. L. 1992. "Role of Genetic Factors in Breast Cancer Susceptibility." *Acta Oncologica* 31 (2): 151–155.

Botkin, Daniel B. 1990. *Discordant Harmonies: A New Ecology for the Twenty-first Century.* New York: Oxford University Press.

——. 1991. "Rethinking the Environment: A New Balance of Nature." *Wilson Quarterly,* Spring: 61–82.

Bramwell, Anna. 1989. *Ecology in the Twentieth Century: A History.* New Haven: Yale University Press.

Brodkin, C. A., S. Barnhart, H. Checkoway, J. Balmes, G. S. Omenn, and L. Rosenstock. 1996. "Longitudinal Pattern of Reported Respiratory Symptoms and Accelerated Ventilatory Loss in Asbestos-Exposed Workers." *Chest* 109, no. 1 (January): 120–126.

Brody, Jane. 1974a. "Liver Cancer Alert Declared to Protect Plastics Workers." *New York Times,* February 15, pp. A1, 66.

——. 1974b. "Cancer Found in Asbestos Workers' Kin." *New York Times,* September 19, pp. A1, 27.

Bronowski, Jacob. 1974. "Technology and Culture in Evolution." In Robert T. Roelofs, Joseph N. Crowley, and Donald L. Hardesty, eds., *Environment and Society: A Book of Readings on Environmental Policy, Attitudes, and Values.* Englewood Cliffs, N.J.: Prentice-Hall.

Brooks, David. 1989. "Journalists and Others for Saving the Planet." *Wall Street Journal,* October 5, p. A28.

Brown, David. 1995. "Weighing the Dangers of a Life-Saver." *Washington Post National Weekly Edition,* May 8: 38.

Budiansky, Stephen. 1995. *Nature's Keepers: The New Science of Nature Management.* New York: Free Press.

Bukro, Casey. 1996. "Clean Air Act Ahead of Schedule; Market Forces Quicken Emissions Reduction." *Chicago Tribune,* March 27: 4.

Burk, Dan L., Kenneth Barovsky, and Gladys H. Munroy. 1993. "Biodiversity and Biotechnology." *Science* 260 (June 25): 1900–1901.

Burns, Michael E. 1988. *Low-Level Radioactive Waste Regulation.* Chelsea, Mich.: Lewis Publishers.

Burros, Marion. 1976. "Peanut Butter: Is Nothing Sacred?" *Washington Post,* July 15, p. E1, E8.

Burroughs, William James. 1997. *Does the Weather Really Matter? The Social Implications of Climate Change*. New York: Cambridge University Press.

Caldwell, Lynton, Lynton R. Hayes, and Isabel M. MacWhirter. 1976. *Citizens and the Environment: Case Studies in Popular Action*. Bloomington: Indiana University Press.

Carson, Rachel. 1962. *Silent Spring*. Boston: Houghton-Mifflin.

Carter, Hodding. 1989. "Alar Scare: Case Study in Media's Skewed Reality." *Wall Street Journal,* April 20, p. A15.

Caulfield, Henry P. 1989. "The Conservation and Environmental Movements: An Historical Analysis." In James P. Lester, ed. *Environmental Politics and Policy: Theories and Evidence*. Durham, N.C.: Duke University Press.

Center for Disease Control (CDC) Veterans Health Study. 1988. "Serum 2,3,7,8-tetrachlorodibenzo-p-dioxin Levels in US Army Vietnam-Era Veterans. *Journal of the American Medical Association* 260:1249–1250.

Center for Risk Analysis. 1995. *Reform of Risk Regulation: Achieving More Protection at Less Cost*. Harvard School of Public Health, March.

Chase, Alston. 1995. *In A Dark Wood*. Boston: Houghton-Mifflin.

Cline, William R. 1992. *Global Warming: The Economic Stakes*. Washington, D.C.: Institute for International Economics.

Coffman, Michael S. 1994. *Saviors of the Earth? The Politics and Religion of the Environmental Movement*. Chicago: Northfield.

Cohen, Aaron J., and C. Arden Pope III. 1995. "Lung Cancer and Air Pollution." *Environmental Health Perspectives* 103, supplement 8 (November): 219–224.

Cohen, Bernard L. 1990. *The Nuclear Energy Option: An Alternative for the '90s*. New York: Plenum Press.

Cole, H. S. D., Marie Jahoda, Christopher Freeman, and K. L. R. Pavitt, eds. 1973. *Models of Doom: A Critique of the Limits of Growth*. New York: Universe Books.

Colglazier, E. W., Jr. 1982. *The Politics of Nuclear Waste*. Elmsford, N.Y.: Pergamon Press.

Commoner, Barry. 1971. *The Closing Circle: Nature, Man, and Technology*. New York: Alfred A. Knopf.

——. 1976. *The Poverty of Power: Energy and the Economic Crisis*. New York: Knopf.

Congressional Quarterly. 1996. Vol. 54, no. 35 (August 31): 2437.

Crandall, Robert W. 1981. *"Pollution Controls and Productivity Growth in Basic Industry" and "Regulation and Productivity Growth": Two Articles by Robert W. Crandall*. Brookings General Series, reprint 375. Washington, D.C.: Brookings Institution.

Cuomo, Christine. 1992. "Unraveling the Problems in Ecofeminism." *Environmental Ethics* 14 (Winter): 351–363.

Curtis, Richard, and Elizabeth Hogan. 1969. *Perils of the Peaceful Atom: The Myth of Safe Nuclear Power.* Garden City, N.J.: Doubleday.

Cushman, John H. 1995. "Republicans Clear-Cut Regulatory Timberland." *New York Times,* March 5, p. A8.

——. 1996a. "E.P.A. Plans Radical Change in Calculation of Cancer Risk." *New York Times,* April 16, pp. A1 and A18.

——. 1996b. "Pesticide Bill Advances in House without Rancor and Opponents." *New York Times,* July 18, pp. A1, A20.

——. 1997. "Environmental Agency Under Fire on Safety Rules." *New York Times,* December 29, p. A16.

——. 1998. "Courts Expanding Efforts to Battle Water Pollution." *New York Times,* March 1, pp. 1, 20.

de Meester, C., and G. B. Gerber. 1995. "The Role of Cooked Food Mutagens as Possible Etiological Agents in Human Cancer: A Critical Appraisal of Recent Epidemiological Investigations." *Revue d'Epidemologie et de Santé Publique* 43 (2): 147–161.

Deneven, William. 1992. "The Pristine Myth: The Landscape of the Americas in 1492." *Annals of the Association of American Geographers* 82 (3): 369–385.

Dennis, Everette E. 1997 "Liberal Reporters, Yes; Liberal Slant, No!" *American Editor* (January/February): 4–9.

Derthick, Martha, and Paul J. Quirk. 1985. *The Politics of Deregulation.* Washington, D.C.: Brookings Institution.

Devesa, Susan S., William J. Blot, B. J. Stone, Barry A. Miller, Robert E. Tarone, and Joseph F. Fraumeni, Jr. 1995. "Recent Cancer Trends in the United States." *Journal of the National Cancer Institute* (February 1): 175–182.

DeVita, Vincent T., Jr., Samuel Helman, and Steven A. Rosenberg, eds. 1985. *Principles and Practice of Oncology.* Philadelphia: Lippincott.

De Witt, Karen. "Challenging the Rule on Additives." *New York Times,* December 10, 1980, pp. C1, C14.

DiLorenzo, Thomas. 1990. *Does Capitalism Cause Pollution?* St. Louis: Center for the Study of American Business, Contemporary Issues Series.

Doble, John. 1995. "Public Opinion about Issues Characterized by Technological Complexity and Scientific Uncertainty." *Public Understanding of Science* 4:95–118.

Doble, John, and Jean Johnson. 1990. *Science and the Public: A Report in Three Volumes.* New York: Public Agenda Forum.

Doll, Richard, and Richard Peto. 1981. *The Causes of Cancer: Quantitative Estimates of Avoidable Risks of Cancer in the U.S. Today.* New York: Oxford University Press.

Doll, Richard, Peter Payne, and John Waterhouse. 1966. *Cancer Incidence in Five Continents.* New York: Springer-Verlag.

Douglas, Mary, and Aaron Wildavsky. 1982. *Risk and Culture: An Essay on the Selection of Technical and Environmental Dangers*. Berkeley and Los Angeles: University of California Press.

Dowie, Mark. 1992. "The New Face of Environmentalism." *Utne Reader,* July/August: 104–111.

Downs, Anthony. 1972. "The Issue Attention Cycle and the Political Economy of Improving Our Environment." In Joe S. Bain and Warren F. Ilchman, eds., *The Political Economy of Environmental Control*. Berkeley, Calif.: Institute of Business and Economic Research.

Dubos, Rene. 1980. *The Wooing of Earth*. New York: Scribner.

Duke, Lynne. 1997. "Limited Trade in Ivory Approved; African Proposal Overrides U.S. Opposition At CITES Conference." *Washington Post,* June 20, p. A16.

Dunlap, Riley E., and Marvin E. Olsen. 1984. "Hard-Path versus Soft-Path Advocates: A Study of Energy Activists." *Policy Studies Journal* 13 (December): 413–428.

Easterbrook, Gregg. 1990. "Everything You Know about the Environment Is Wrong." *New Republic,* April 30: 14–27.

——. 1994. "The Birds." *New Republic,* March 28: 22–23, 27–29.

——. 1995. *A Moment on the Earth: The Coming Age of Environmental Optimism*. New York: Viking.

Economist. 1995. "Greenpeace Means Business." August 19: 59–62.

——. 1996a. "The Cigarette Wars." May 11: 21–23.

——. 1996b. "Overregulating America." July 27: 19–21.

——. 1997. "Clean Air, Dirty Fight." March 15: 29–30.

——. 1998. "The Earth's Climate: Solar Rash and Earthly Fever." February 21, pp. 81–82.

Eeles, R. A., M. R. Stratton, D. E. Goldgar, and D. F. Easton. 1994. "The Genetics of Familial Breast Cancer and Their Practical Implications." *European Journal of Cancer*. 30A, 9: 1383–1390.

Efron, Edith. 1984. *The Apocalyptics: Cancer and the Big Lie*. New York: Simon and Schuster.

Ehrlich, Paul R. 1968. *The Population Bomb*. New York: Ballantine.

Ehrlich, Paul R., and Anne H. Ehrlich. 1996. *Betrayal of Science and Reason*. Washington, D.C.: Island Press.

Eisenbud, Merril. 1973. *Environmental Radioactivity*. 2d ed. New York: Academic Press.

——. 1978. *Environment, Technology, and Health: Human Ecology in Historical Perspective*. New York: New York University Press.

Encyclopedia of Associations, 1994. 1993. 28th ed. Detroit: Gale Research.

Environment. 1972. Vol. 14 (April).

Faich, Ronald O., and Richard P. Gale. 1971. "The Environmental Movement: From Recreation to Politics." *Pacific Sociological Review* 14 (3): 270–287.

Faltermayer, Edmund. 1988. "Taking Fear out of Nuclear Power." *Fortune,* August 1: 105–118.

Ferry, Luc. 1995. *The New Ecological Order.* Translated by Carol Volk. Chicago: University of Chicago Press.

Feshbach, Murray, and Alfred Friendly, Jr. 1992. *Ecocide in the USSR: Health and Nature Under Siege.* New York: Basic Books.

Finkel, Adam. 1996a. "Who's Exaggerating?" *Discover,* May: 48–54.

——. 1996b. "A Return to Alchemy." *Environmental Forum,* Sept/Oct: 15–19.

Firor, John. 1990. *The Changing Atmosphere: A Global Challenge.* New Haven: Yale University Press.

Fischer, David S. 1982. "The Etiologies of Cancer." In David S. Fischer, ed., *Cancer Therapy.* Boston: G. K. Hall.

Fischoff, Baruch. 1985. "Managing Risk Perception." *Issues in Science and Technology,* Fall: 83–96.

Fisher, David E. 1990. *Fire and Ice: The Greenhouse Effect, Ozone Depletion, and Nuclear Winter.* New York: Harper and Row.

Fontham, E. T., P. Correa, V. W. Chen. 1993. "Passive Smoking and Lung Cancer." *Journal of Louisiana St Medical Society* 145, 4 (Apr): 132–136.

Fox, Stephen. 1981. *John Muir and His Legacy: The American Conservation Movement.* Boston: Little, Brown.

Freedman, Allan. 1997. "Congress Prepares New Assault on Troubled Superfund Sites." *Congressional Quarterly,* June 28: 1502–1507.

——. 1998. "EPA's Reversal on Herbicide Ban Sows Concern and Confusion," *CQ Weekly,* May 2, p. 1118.

Freeman, J. Leiper. 1955. *The Political Process: Executive Bureau-Legislative Committee Relations.* Rev. ed. 1965. New York: Random House.

Frischler, F. Lee. 1989. *Smoking and Politics.* 4th ed. Englewood Cliffs, N.J.: Prentice-Hall.

Fuchs, C. S., and R. J. Mayer. 1995. "Gastric Carcinoma." *New England Journal of Medicine* 333 (1): 32–41.

Fumento, Michael. 1993. "Mother Nature's Pesticide Factory." *New York Times,* June 9: 21.

——. 1997. *Polluted Science.* Washington D.C.: AEI Press.

Gale, Richard P. 1983. "The Environmental Movement and the Left: Antagonists or Allies?" *Sociological Inquiry* 53 (Spring/Summer): 179–199.

Geertz, Clifford. 1973. "Ideology as a Cultural System." In Clifford Geertz, *The Interpretation of Cultures.* New York: Basic Books.

Gerrans, Phillip. 1997. *Times Literary Supplement,* June 20: 4–5.

Gillette, Clayton P., and James E. Krier. 1990. "Risks Courts and Agencies." *University of Pennsylvania Law Review* 138 (April): 1027–1109.

Giusti, R. M., K. Iwanmoto, and E. E. Hatch. 1995. "Diethylstilbestrol Revisited: A Review of the Long-Term Health Effects." *Annals of Internal Medicine* 122 (10): 778–788.

Glasser, Ronald J. 1976. "The Real Cause of Cancer." *Washington Monthly,* October: 4–17.

Gofman, John W., and Arthur R. Tamplin. 1971. *Poisoned Power: The Case against Nuclear Power Plants*. Emmaus, Penn.: Rodale Press.

Gomer, Robert. 1974. "The Tyranny of Progress." In Robert T. Roelofs, Joseph N. Crowley, and Donald L. Hardesty, eds. *Environment and Society: A Book of Readings on Environmental Policy, Attitudes, and Values*. Englewood Cliffs, N.J.: Prentice-Hall.

Goodfield, June. 1981. *Reflections on Science and the Media*. New York: American Association for the Advancement of Science.

Gordon, John Steele. 1993. "The American Environment." *American Heritage* 44 (6): 30–51.

Gori, Gio B. 1980. "The Regulation of Carcinogenic Hazards." *Science* 208 (April 18): 256–261.

Gottlieb, Robert. 1991. "An Odd Assortment of Allies: American Environmentalism in the 1990s." In Craig L. LaMay and Everette E. Dennis, eds., *Media and the Environment*. Washington, D.C.: Island Press.

Gough, Michael. 1989. "Estimating Cancer Mortality." *Environmental Science and Technology* 23:925–930.

——. 1990. "How Much Cancer Can the EPA Regulate Anyway?" *Risk Analysis* 10:1–6.

——. 1991. "Agent Orange: Exposure and Policy." *American Journal of Public Health* 81:289–290.

Graham, John. 1995. "Making Sense of Risk." Unpublished paper presented by the Center for Risk Analysis, Harvard School of Public Health, Boston, Mass.

Graham, John, and Jonathan Weiner. 1995. *Risk versus Risk*. Cambridge, Mass.: Harvard University Press.

Gregg, Frank. 1991. "Public Land Policy: Controversial Beginnings for the Third Century." In Michael J. Lacey, ed. *Government and Environmental Politics*. Washington, D.C.: Woodrow Wilson Center Press.

Gross, Paul, and Norman Levitt. 1994. *Higher Superstition: The Academic Left and Its Quarrels with Science*. Baltimore: Johns Hopkins University Press.

Gross, Paul, Norman Levitt, and Martin W. Lewis, eds. 1996. *The Flight From Science and Reason*. New York: New York Academy of Sciences.

Harris, Robert H., Robert B. Nicholas, and Paul Milvy. 1981. "Reducing Environmental Risks." *Society,* March/April: 17–19.

Harrison, Kathryn, and George Hoberg. 1994. *Risk, Science and Politics*. Montreal: McGill-Queen's University Press.

Hartz, Jim, and Rick Chappell. 1997. *Worlds Apart: How the Distance Between Science and Journalism Threatens America's Future*. Nashville: First Amendment Center.

Harvard Center for Risk Analysis. 1996. *Risk in Perspective* 4 (4): 1–4.

Harvard Group on Risk Management Reform. 1995. *Reform of Risk Regulation: Achieving More Protection at Low Cost.* Boston: Center for Risk Analysis, Harvard School of Public Health.

Harvard Health Letter. 1998. "Medical Progress: New Hope for Prevention, Treatment of Breast Cancer." (August), p. 6.

Havender, William R. 1983a. "Conned by Pros: Dishonest Science and the Public Policy of Risk." Berkeley: University of California, Berkeley, Department of Biochemistry.

——. 1983b. "The Science & Politics of Cyclamate." *Public Interest,* Sp: 17–32.

Hays, Samuel P. 1959. *Conservation and the Gospel of Efficiency: The Progressive Conservation Movement, 1890–1920.* Cambridge: Harvard University Press.

——. 1987. *Beauty, Health, and Permanence: Environmental Politics in the United States, 1955–1985.* Cambridge: Cambridge University Press.

Heclo, Hugh. 1978. "Issue Networks and the Executive Establishment." In Anthony King, ed., *The New American Political System.* Washington, D.C.: American Enterprise Institute.

Hellman, Samuel, and Steven A. Rosenberg, eds. 1982. *Cancer: Principles and Practice of Oncology.* Philadelphia: Lippincott.

Hertsgaard, Mark. 1996. "A World Awash in Chemicals." *New York Times Book Review,* April 7, 1996: 25.

Higginson, John, and C. S. Muir. 1979. "Environmental Carcinogenesis: Misconceptions and Limitations to Cancer Control." *Journal of the National Cancer Institute* 63 (6): 1291–1298.

Hilts, Philip J. *Smokescreen: The Truth behind the Tobacco Industry Cover-Up.* Reading, Mass.: Addison-Wesley.

Hooper, Joseph. 1990. "The Asbestos Mess." *New York Times Magazine,* November 25: 38ff.

Houben, G. F., P. M. Abma, H. van den Berg, W. van Dokkum, H. van Loveren, A. H. Penninks, W. Seinen, S. Spanhaak, J. G. Vos, and T. Ockhuizen. 1992. "Effects of the Color Additive Caramel Color III on the Immune System: A Study with Human Volunteers." *Food and Chemical Toxicology* 30 (9): 749–757.

Houghton, John. 1997. *Global Warming: The Complete Briefing.* 2d ed. New York: Cambridge University Press.

Huber, Peter W. 1991. *Galileo's Revenge: Junk Science in the Courtroom.* New York: Basic Books.

Hueper, Wilhelm C. 1942. *Occupational Tumors and Allied Diseases.* Springfield, Ill.: Thomas.

Huth, Hans. 1957. *Nature and the American: Three Centuries of Changing Attitudes.* Berkeley and Los Angeles: University of California Press.

Hypatia. 1991. Special edition, 6 (Winter).

Inglehart, Ronald. 1977. *The Silent Revolution: Changing Values and Political Styles among Western Publics*. Princeton, N.J.: Princeton University Press.

——. 1990. *Culture Shift in Advanced Industrial Society*. Princeton, N.J.: Princeton University Press.

——. 1995. "Public Support for Environmental Protection." *PS* 28 (1): 57–71.

Ingram, Helen M., and Dean E. Mann. 1984. "Preserving the Clean Water Act: The Appearance of Environmental Victory." In Norman J. Vig and Michael E. Kraft, eds., *Environmental Policy in the 1980s: Reagan's New Agenda*. Washington, D.C.: Congressional Quarterly Press.

Institute of Medicine, 1993. *Veterans and Agent Orange*. Washington, D.C.: National Academy Press.

Jackson, James. 1992. "Nuclear Time Bombs." *Time,* December 7: 45.

Jasper, James M. 1990. *Nuclear Politics: Energy and the State in the United States, Sweden, and France*. Princeton, N.J.: Princeton University Press.

Jukes, Thomas. 1993. "Pesticide Residues in Food." *Priorities,* Summer: 7–9.

Kelman, Steven. 1980. "Occupational Safety and Health Administration." In James Q. Wilson, ed., *The Politics of Regulation*. New York: Basic Books.

——. 1981. "Economists and the Environmental Muddle." *Public Interest,* Summer: 106–117.

Keniry, Julian. 1993. "Environmental Movement Booming on Campuses." *Change,* September/October: 42–49.

Kenski, Henry C., and Helen M. Ingram. 1986. "The Reagan Administration and Environmental Regulation: The Constraint of the Political Market." In Sheldon Kamieniecki, Robert O'Brien, and Michael Clarke, eds., *Controversies in Environmental Policy*. Albany: State University of New York Press.

Kerr, Richard A. 1995. "Scientists See Greenhouse, Semiofficially." *Science* 269 (September 22): 1667.

Klaidman, Stephen. 1991. "Muddling Through." *Wilson Quarterly* (Spring): 73–82.

Klugar, Richard. 1996. *Ashes to Ashes: America's Hundred-Year Cigarette War, the Public Health, and the Unabashed Triumph of Philip Morris*. New York: Knopf.

Knox, Margaret. 1992. "The Grass-Roots Anti-Environmental Movement." *Utne Reader,* July/August: 108–109.

Kolata, Gina. 1995a. "A Case of Justice, or a Total Travesty?" *New York Times,* June 13, pp. D1, D5.

——. 1995b. "Proof of a Breast Implant Peril Is Lacking, Rheumatologists Say." *New York Times,* October 25, p. C11.

——. 1996a. "Chemicals That Mimic Hormones Spark Alarm and Debate." *New York Times,* March 19, 1996, pp. B5, B10.

——. 1996b. "Sperm Counts: Some Experts See a Fall, Others Poor Data." *New York Times,* March 19, 1996, p. B10.

——. 1996c. "New System of Assessing Health Risks Is Urged." *New York Times,* June 14, p. A26.

——. 1997a. "Key Study Sees No Evidence Power Lines Cause Leukemia." *New York Times,* July 3, pp. A1, A23.

——. 1997b. "Study Discounts DDT Role in Breast Cancer." *New York Times,* October 30, p. A26.

Kuhn, Thomas. 1970. *The Structure of Scientific Revolutions*. 2d, enlarged ed. Chicago: University of Chicago Press.

Kuipers, Dean. 1992. "Pirates with a Difference: An Interview with Paul Watson." *Interview* 22 (August): 94–95.

Kurtz, Howard. 1993. "Stories on Cancer Causes Are Said to Be Misfocused." *Washington Post National Weekly Edition,* July 27, p. A6.

——. 1994. "The Great Exploding Popcorn Exposé." *Washington Post,* December 5, p. C1.

Ladd, Everett Carll, and Karlyn H. Bowman. 1995. *Attitudes toward the Environment*. Washington, D.C.: American Enterprise Institute.

Landy, Marc K., and Mary Hague, 1992. "Private Interests and Superfund." *Public Interest* 108 (Summer): 97–115.

Landy, Marc K., Marc J. Roberts, and Stephen R. Thomas. 1990. *The Environmental Protection Agency: Asking the Wrong Questions*. New York: Oxford University Press.

Lave, Lester B. 1992. "Regulation." *American Enterprise* 3, no. 6 (November/December): 19–22.

Lave, Lester B., and Arthur C. Upton. 1987. *Toxic Chemicals, Health, and the Environment*. Baltimore, Md.: Johns Hopkins University Press.

Lee, Gary. 1994. "A Potential Killer's Modus Operandi: An Upcoming EPA Report Outlines the Effects of Dioxin on Humans." *Washington Post National Weekly Edition,* June 20–26: 38.

Lehman, Stephen. 1990. "Greenpeace." *Organization Trends,* July: 1–8.

Lerner, Robert, Althea Nagai, and Stanley Rothman. 1996. *American Elites*. New Haven: Yale University Press.

Lewis, Martin W. 1992. *Green Delusions: An Environmentalist Critique of Radical Environmentalism*. Durham, N.C.: Duke University Press.

Lichter, Linda S., S. Robert Lichter, and Stanley Rothman 1983. "What Interests the Public and What Interests the Public Interests." *Public Opinion* 6, no. 2 (April–May): 44–48.

Lichter, S. Robert, 1996. "Consistently Liberal: But Does It Matter?" *Forbes Media Critic,* (Fall): pp. 26–39.

Lichter, S. Robert. 1993. "Why Cancer News Is a Health Hazard." *Wall Street Journal,* November 12, p. A14.

Lichter, S. Robert, Stanley Rothman, and Linda S. Lichter. 1986. *The Media Elite: America's New Powerbrokers*. Washington, D.C.: Adler & Adler.

Lin, C. S. 1992. "Evaluating the Safety of Food and Color Additives with Pharmacokinetic Data." 32 (2):191–195.

Litan, Robert, and Clifford Winston, eds. 1988. *Liability: Perspectives and Policy.* Washington, D.C: Brookings Institution.

Lovelock, James. 1990. *The Ages of Gaia.* New York: Bantam Books.

Low, Bobbi S. 1996. "Behavioral Ecology of Conservation in Traditional Societies." *Human Nature* 7 (4): 353–379.

Lunch, William. 1987. *The Nationalization of American Politics.* Berkeley and Los Angeles: University of California Press.

Mahlin, Michelle M. 1995. "Healing Power: Low-Level Radiation in Perspective." *Priorities* 7 (2): 10–12.

Mahoney, F. X., Jr. 1992. "Saccharin's Link to Human Cancer Questioned." *Journal of the National Cancer Institute* 84 (9): 665–667.

Mahoney, James R. 1990. "National Acid Rain Precipitation Assessment Program—Key Results." Paper presented at the NAPAP Final Task Force Meeting, September 5, National Academy of Sciences, Washington, D.C.

Malkin, D., and C. Portwine. 1994. "The Genetics of Childhood Cancer." *European Journal of Cancer* 30A (13): 1942–1946.

Manes, Christopher. 1990. *Green Rage: Radical Environmentalism and the Unmaking of Civilization.* Boston: Little, Brown.

Mann, Charles, and Mark Plummer. 1995. *Noah's Choice: The Future of Endangered Species.* New York: Knopf.

Margolis, Howard. 1996. *Experts, Victims, and Risk.* Chicago: University of Chicago Press.

Marsh, George Perkins. 1965. *Man and Nature* (1864). Cambridge, Mass.: Belknap Press at Harvard University Press.

Marshall, Eliot. 1991. "A Is for Apple, Alar, and . . . Alarmist?" *Science* 254 (October 4): 20–22.

——. 1993. "Is Environmental Technology a Key to a Healthy Economy?" *Science* 260 (June 25): 1884–1888.

Masood, Ehsan. 1995. "Climate Panel Confirms Human Role in Warming, Fights Off Oil States." *Nature* 378 (December 7): 524.

Mazur, Allan. 1981. *The Dynamics of Technical Controversy.* Washington, D.C.: Communications Press.

McConnell, Grant. 1954. *The Conservation Movement: Past and Present.* Salk Lake City: Institute of Government, University of Utah.

McCormick, John. 1989. *Reclaiming Paradise: The Global Environmental Movement.* Bloomington: Indiana University Press.

Meadows, Donella H., Dennis L. Meadows, Jorgen Randers, and William W. Behrens III. 1972. *The Limits to Growth: A Report of the Club of Rome's Project on the Predicament of Mankind.* New York: Universe Books.

Media Monitor. 1987. "The AIDS Story: Science, Politics, Sex and Death." Vol. 1, no. 9 (December): Whole Issue.

Melnick, R. Shep. 1983. *Regulation and the Courts: The Case of the Clean Air Act.* Washington, D.C.: Brookings Institution.

Melosi, Martin V. 1992. "Pollution and the Emergence of Industrial America." In Judith E. Jacobsen and John Firor, eds., *Human Impact on the Environment: Ancient Roots, Current Challenges.* Boulder, Colo.: Westview Press.

Michaels, Patrick J. 1997. "Holes in the Greenhouse Effect?" *Washington Post National Weekly Edition,* June 30: 23.

Milbrath, Lester. 1984. *Environmentalists: Vanguard for a New Society.* Albany: State University of New York Press.

Miller, G. H., J. A. Golish, C. E. Cox, and D. C. Chacko. 1994. "Women and Lung Cancer: A Comparison of Active and Passive Smokers with Nonexposed Nonsmokers." *Cancer Detection and Prevention* 18 (6): 421–430.

Miller, R. W. 1995. "Delayed Effects of External Radiation Exposure: A Brief History." *Radiation Research* 144 (2): 160–169.

Milloy, Steven J., and Michael Gough. 1997. "The EPA's Clean Air-ogance." *Wall Street Journal,* January 7, p. A16.

Mishell, D. R., Jr. 1994. "Breast Cancer Risk with Oral Contraceptives and Oestrogen Replacement Therapy." *Australian and New Zealand Journal of Obstetrics and Gynaecology* 34 (3): 316–320.

Mitchell, Robert C. 1991. "From Conservation to Environmental Movement: The Development of the Modern Environmental Lobbies." In Michael J. Lacey, ed., *Government and Environmental Politics.* Washington, D.C.: Woodrow Wilson Center Press.

Mitchener, Brandon. 1991. "Out on a Limb for Mother Earth." *The World and I,* July: 573–589.

Mittelbach, Margaret, and Mitchell Crewdson. 1994. "Botanical Survey of Long Island park Finds Plants That Might Have Medicinal Value." *Newsday,* December 6, p. B29.

Moolgavkar, Suresh H., and E. Georg Luebeck. 1996. "A Critical View of the Evidence on Particulate Air Pollution and Mortality." *Epidemiology* 103 (November): 219–224.

Moolgavkar, Suresh H., E. Georg Luebeck, Thomas A. Hall, and Elizabeth L. Anderson. 1995. "Air Pollution and Daily Mortality in Philadelphia." *Epidemiology* 6 (5): 476–484.

Moore, John A. 1987. "The Not So Silent Spring." In Gino J. Marco, Robert M. Hollingworth, and William Durham, eds., *Silent Spring Revisited.* Washington, D.C.: American Chemical Society.

Morris, P. D. 1995. "Lifetime Excess Risk of Death from Lung Cancer for a U.S. Female Never-Smoker Exposed to Environmental Tobacco Smoke." *Environmental Research* 68, no. 1 (January): 3–9.

Moss, Ralph W. 1980. *The Cancer Syndrome.* New York: Grove Press.

Mumford, Lewis.1953. *The Highway and the City.* New York: Harcourt Brace.

Murray, David. 1996a. "Global Climate Change." *Stats 2100: Newsletter of the Statistical Assessment Service* 2, no. 2 (July): 3–9.

——. 1996b. "The Sperm Count's Falling! The Sperm Count's Falling!" *Stats 2100: Newsletter of the Statistical Assessment Service* 2, no. 2 (July): 1–2.

Nash, Roderick. 1976. *The American Environment: Readings in the History of Conservation*. 2d ed. Reading, Mass.: Addison-Wesley.

——. 1982. *Wilderness and the American Mind*. 3d ed. New Haven: Yale University Press.

——. 1989. *The Rights of Nature: A History of Environmental Ethics*. Madison: University of Wisconsin Press.

National Research Council. 1993. *Pesticides in the Diet of Infants and Children*. Washington, D.C. National Academy Press.

——. 1996. *Carcinogens and Anti-Carcinogens in the Human Diet*. Washington, D.C.: National Academy Press.

Nelkin, Dorothy. 1985. "Managing Biomedical News." *Social Research* 52: 625–646.

Nelson, Robert H. 1990. "Tom Hayden, Meet Adam Smith and Thomas Aquinas." *Forbes,* October 29: 94–97.

——. 1996. "Salem Revisited." *Forbes,* November 4: 49.

Newell, Guy R., W. Bryant Boutwell, Dexter L. Morris, Barbara C. Tilley, and Evelyn S. Branyon. 1982. "Epidemiology of Cancer." In Vincent T. DeVita, Jr., Samuel Hellman, and Steven A. Rosenberg, eds., *Cancer: Principles and Practice of Oncology*. Philadelphia: Lippincott.

New York Times. 1975. "Additional Evidence Points to Daily Diet as a Cancer Cause." December 3: 1, 79.

——. 1978. "Upstate Waste Site May Endanger Lives—Abandoned Dump in Niagara Falls Leaks Possible Carcinogens." August 2, pp. A1, B9.

——. 1984. "After Years of Cancer Alarms, Progress amid the Mistakes." March 20, pp. C1, C12, C13.

——. 1995. "Cancer Fear Is Unfounded, Physicists Say: Power Line Concern Is Called Needless." May 14: 19.

——. 1996a. "Fine Pollutants in Air Cause Many Deaths." May 9, p. A8.

——. 1996b. "Microscopic Killers." May 12, p. E12.

——. 1996c. "Americans Flunk Science, A Study Shows." May 24, p. A12.

——. 1997a. "Clinton Defers Curbs on Gases Heating Globe." June 27, pp. A1, A11.

——. 1997b. "Pesticide Risk from Produce Is Called Slight." Nov. 16, p. 32.

Nichols, Mark. 1998. "An Ounce of Prevention: Tests Show A Drug Can Stop Breast Cancer from Starting," *Maclean's,* April 20, pp. 50–51.

Nicholson, Max. 1987. *The New Environmental Age*. Cambridge, Mass.: Cambridge University Press.

Nivola, Pietro. 1996. "Having It All?" *Brookings Review,* Winter: 14–21.

Novick, Sheldon N. 1969. *The Careless Atom*. Boston: Houghton-Mifflin.

Occupational Safety and Health Administration. 1978. "Estimates of the Fraction of Cancer in the United States Related to Occupational Factors." Washington, D.C.: NCI/NIEHS/NIOSH.

Oelschlaeger, Max. 1991. *The Idea of Wilderness: From Prehistory to the Age of Ecology*. New Haven: Yale University Press.

Office of Technology Assessment. 1981. *Cancer Risk: Assessing and Reducing the Dangers in Our Society*. Washington, D.C.: Government Printing Office.

Oregonian. 1995. "Public Tells Pollsters That GOP Congress Doesn't Make Grade." April 7, p. A14.

O'Riordan, Timothy. 1981. *Environmentalism*. London: Pion.

Ottoboni, M. Alice. 1984. *The Dose Makes the Poison*. Berkeley, Calif.: Vincente.

Paehlke, Robert C. 1989. *Environmentalism and the Future of Progressive Politics*. New Haven: Yale University Press.

Palmer, J. R., L. Rosenberg, R. S. Rao, B. L. Strom, M. E. Warshauer, S. Harlap, A. Zauber, and S. Shapiro. 1995. "Oral Contraceptive Use and Breast Cancer Risk among African-American Women." *Cancer Causes and Control* 6 (4): 321–331.

Palmer, Karen, Wallace E. Oates, and Paul Portney. 1995. "Tightening Environmental Standards: The Benefit-Cost or the No-Cost Paradigm?" *The Journal of Economic Perspectives*, vol. 9, no. 4: 119.

Parrish, Michael. 1993. "Capitalist Cleanup." *Los Angeles Times*, March 31, p. D1.

Passell, Peter. 1991a. "The Garbage Problem: It May Be Politics, Not Nature." *New York Times*, February 26, pp. C1, C6.

——. 1991b. "Experts Question Staggering Costs of Toxic Cleanups." *New York Times*, September 1: 28.

Patterson, James T. 1987. *The Dread Disease: Cancer and Modern American Culture*. Cambridge, Mass.: Harvard University Press.

Pearce, Fred. 1991. *Green Warriors: The People and the Politics behind the Environmental Revolution*. London: Bodley Head.

Pearl, Mary. 1995. "Asian Rhinos and Elephants Go Quietly." *New York Times*, April 19, p. A22.

Peto J., K. F. Easton, F. E. Matthews, D. Ford, and A. J. Swerdlow. 1996. "Cancer Mortality in Relatives of Women with Breast Cancer: The OPCS Study." Office of Population Censuses and Surveys. *International Journal of Cancer* 65, no. 3 (January): 275–283.

Petulla, Joseph M. 1977. *American Environmental History: The Exploitation and Conservation of Natural Resources*. San Francisco: Boyd & Fraser.

——. 1980. *American Environmentalism: Values, Tactics, Priorities*. College Station: Texas A & M University Press.

Pope, C. Arden, III, Michael J. Thun, Mohan M. Namboodiri, Douglas W. Dockery, John S. Evans, Frank E. Speizer, and Clark W. Heath, Jr. 1995. "Particulate Air Pollution as a Predictor of Mortality in Prospective Study of U.S. Adults." *American Journal of Respiratory and Critical Care Medicine* 151, no. 3 (March): 669–674.

Pope, Harrison G., Jr., and S. Robert Lichter. 1989. "The Reporting of AIDS." Letter to the editor. *Journal of American Medical Association* 262, no. 14 (October 13): 1949–1950.

Porter, Michael E., and Claas van der Linde. 1995. "Toward a New Conception of the Environment-Competitiveness Relationship." *Journal of Economic Perspectives,* vol. 9, no. 4: 97.

Potter, David M. 1954. *People of Plenty: Economic Abundance and the American Character.* Chicago: University of Chicago Press.

Price, Don K. 1962. "The Scientific Establishment." *Proceedings of the American Philosophical Society,* June: 70–91.

——. 1965. *The Scientific Estate.* Cambridge: Harvard University Press.

Probst, Katherine N., Don Fullerton, Robert E. Litan, and Paul R. Portney. 1995. *Footing the Bill for Superfund Cleanups.* Washington, D.C.: Brookings Institution and Resources for the Future.

Proctor, Robert N. 1995. *Cancer Wars: How Politics Shapes What We Know and Don't Know about Cancer.* New York: Basic Books.

Purdum, Todd S. 1996. "Clinton Orders Expanded Agent Orange Benefits for Veterans." *New York Times,* May 29: 17.

Rathje, William L., and Cullen Murphy. 1992. *Rubbish!: The Archeology of Garbage.* New York: HarperCollins.

Ray, Dixie Lee, with Lou Guzzo. 1990. *Trashing the Planet: How Science Can Help Us Deal with Acid Rain, Depletion of the Ozone, and Nuclear Waste (among Other Things).* Washington, D.C.: Regnery Gateway.

Renke, W., and E. Rosik. 1994. "Distant Health Effects of Using Asbestos in Shipyards and in Co-operating Plants." *Bulletin of the Institute of Maritime and Tropical Medicine in Gdynia* 44–45, nos. 1–4: 5–11.

Rensberger, Boyce. 1993. "Blowing Hot and Cold on Global Warming." *Washington Post National Weekly Edition,* August 2–8: 38.

Rhomberg, Lorenz, and Sonia Hernandez Diaz. 1997. "The Sperm Count Debate." *Risk in Perspective* 5, no. 11 (November): 1–4.

Richards, Victor. 1972. *The Wayward Cell.* Berkeley and Los Angeles: University of California Press.

Rosen, Joseph D. 1990. "Much Ado about Alar." *Issues in Science and Technology,* Fall: 85–90.

Rosenbaum, Walter A. 1991. *Environmental Politics and Policy.* 2d ed. Washington, D.C.: Congressional Quarterly Press.

Rothman, Kenneth, and Andrew Keller. 1972. "The Effect of Joint Exposure to

Alcohol and Tobacco on the Risk of Cancer of the Mouth and Pharynx." *Journal of Chronic Diseases* 25: 711–716.

Rothman, Stanley. 1994. Panel 1, "Liberty, Property, and Environmental Ethics." *Ecology Law Quarterly* 21, no. 1: 387–429.

Rothman, Stanley, ed. 1992. *The Mass Media in Liberal Democratic Societies.* New York: Paragon House.

Rothman, Stanley, and Amy Black. Forthcoming. *American Elites Revisited.*

Rothman, Stanley, and S. Robert Lichter. 1987. "Elite Ideology and Risk Perception in Nuclear Energy Policy." *American Political Science Review* 81 (June): 383–404.

Rourke, Francis E. 1984. *Bureaucracy, Politics, and Public Policy*. 3d ed. Boston: Little, Brown.

Rubin, Charles T. 1989. "Environmental Policy and Environmental Thought: Ruckelshaus and Commoner." *Environmental Ethics* 11 (Spring): 27–51.

———. 1994. *The Green Crusade: Rethinking the Roots of Environmentalism.* New York: Free Press.

Ruether, Rosemary Radford. 1992. *Gaia and God: An Ecofeminist Theology of Earth Healing*. San Francisco: HarperCollins.

Rushefshy, Mark E. 1986. *Making Cancer Policy*. Albany: State University of New York Press.

Russell, Christine. 1990. "As American as Alar Pie." *Washington Post National Weekly Edition,* March 19: 6–7.

Sale, Kirkpatrick. 1993. "The U.S. Green Movement Today." *The Nation,* July 19: 92–96.

Salleh, Ariel. 1992. "The Ecofeminism/Deep Ecology Debate: A Reply to Patriarchal Reason." *Environmental Ethics* 14 (Fall): 195–216.

Salsburg, David, and Andrew Heath. 1981. "When Science Progresses and Bureaucracies Lag." *Public Interest* 65 (Fall): 30–39.

Samet, Jonathan M. 1996. "Small Particles, High Stakes." *Johns Hopkins University School of Hygiene and Public Health,* Fall: 1–5.

Samet, Jonathan M., and Aaron J. Cohen. 1995. "Air Pollution and Lung Cancer." Unpublished Draft Manuscript, August 17.

Samuelson, Robert J. 1996. *The Good Life and Its Discontents: The American Dream in the Age of Entitlement.* New York: Times Books.

Scanlon, Kevin J. Mohammed Kashani-Sabet. "Ribozymes as Therapeutic Agents: Are We Getting Closer?" *Journal National Cancer Institute,* vol. 90: 558–559.

Schapira, D. V. 1992. "Nutrition and Cancer Prevention." *Primary Care Clinics in Office Practice* 19, no. 3: 481–491.

Scheffer, Victor B. 1991. *The Shaping of Environmentalism in America.* Seattle: University of Washington Press.

Schmeck, Jr., Harold M. 1957. "Cigarette Smoking Linked To Cancer in High Degree." *New York Times,* June 5: 1, 25.

Schneider, Keith. 1992. "Panel of Scientists Finds Dioxin Does Not Pose Widespread Cancer Threat." *New York Times,* September 26, p. 9.

——. 1993a. "New View Calls Environmental Policy Misguided." *New York Times,* March 21, pp. A1, A30.

——. 1993b. "How a Rebellion over Environmental Rules Grew from a Patch of Weeds." *New York Times,* March 24, p. A16.

——. 1993c. "Second Chance on Environment." *New York Times,* March 26, p. A17.

——. 1994a. "Fetal Harm, Not Cancer, Is Called the Primary Threat from Dioxin." *New York Times,* May 11, pp. A1, A20.

——. 1994b. "EPA Moves to Reduce Health Risks from Dioxin." *New York Times,* September 14: 15.

Schneider, Stephen H., 1988. *Global Warning: Are We Entering the Greenhouse Century?* San Francisco: Sierra Club Books.

Schneider, William, and I. A. Lewis. 1985. "Views on the News." *Public Opinion* 8, no. 4 (August/September): 6–11, 58–59.

Schroeder, Christopher. 1991. "The Evolution of Federal Regulation of Toxic Substances." In Michael J. Lacey, ed., *Government and Environmental Politics.* Washington, D.C.: Woodrow Wilson Center Press.

Schuck, Peter. 1986. *Agent Orange on Trial.* Cambridge, Mass.: Harvard University Press.

Schudson, Michael. 1995. *The Power of News.* Cambridge, Mass.: Harvard University Press.

Science. 1995a. Vol. 269 (May 26): 1124.

——. 1995b. Vol. 269 (September 24): 1667.

Science News. 1996. Vol. 149, June 8: 365.

SEER Cancer Statistics Review, 1973–1991: Tables and Graphs. 1994. Bethesda, Md.: U.S. Department of Health and Human Services, Public Health Service, National Institutes of Health.

Sessions, Robert. 1991. "Deep Ecology versus Ecofeminism: Healthy Differences or Incompatible Philosophies?" *Hypatia* 6 (Spring): 90–107.

Shabecoff, Phillip. 1988. "EPA Reassesses the Cancer Risks of Many Chemicals." *New York Times,* January 4, pp. A1, A17.

Sharpe, R. M., and N. E. Skakkebaek. 1993. "Are Oestrogens Involved in Falling Sperm Counts and Disorders of the Male Reproductive Tract?" *Lancet* 341, no. 8857: 1392–1395.

Shaw, David. 1994a. "Headlines and High Anxiety." *Los Angeles Times,* September 11, pp. A1, A30–A32.

——. 1994b. "Controversial Stories Go against the Grain." *Los Angeles Times,* September 11, p. A31.

——. 1994c. "Dose of Skepticism Enters Coverage of the Environment." *Los Angeles Times,* September 11, p. A32.

——. 1994d. " 'Cry Wolf' Stories Permeate Coverage of Health Hazards." *Los Angeles Times,* September 12, p. A1.

——. 1994e. "Alar Panic Shows Power of Media to Trigger Fear." *Los Angeles Times,* September 12, p. A19.

——. 1994f. "Media Speak to a Public Ripe to Find Health Danger." *Los Angeles Times,* September 13, p. A1.

——. 1994g. "Does Rise in Breast Cancer Cases Constitute Epidemic?" *Los Angeles Times,* September 13, p. A16.

——. 1994h. "Nuclear Power Coverage Focused Morbidly on Risk." *Los Angeles Times,* September 13, p. A17.

Siegel, M., 1993. "Involuntary Smoking in the Restaurant Workplace: A Review of Employee Exposure and Health Effects." *Journal of the American Medical Association* 270, no. 4 (July 28): 490–493.

Simon, Julian L. 1981. *The Ultimate Resource.* Princeton, N.J.: Princeton University Press.

Slovic, Paul. 1987. "Perception of Risk." *Science,* April 17: 280–285.

Smith, Kenneth. 1994. "Labels Show the Media's Bias." *Priorities* (6): 1, 38–39.

Smith, Zachary A. 1992. *The Environmental Policy Paradox.* Englewood Cliffs, N.J.: Prentice Hall.

Snow, Donald. 1992. *Inside the Environmental Movement: Meeting the Leadership Challenge.* Washington, D.C.: Island Press.

Specter, Michael. 1993. "Sea-Dumping Ban: Good Politics, but Not Necessarily Good Policy." *New York Times,* March 22, pp. A1, B8.

Speth, James Gustave. 1992. "Transitions to a Sustainable Society." In Judith E. Jacobsen and John Firor, eds., *Human Impact on the Environment: Ancient Roots, Current Challenges.* Boulder, Colo.: Westview Press.

Stanfield, Rochelle L. 1985. "Environmental Lobby's Changing of the Guard Is Part of Movement's Evolution: The Incoming Leaders Are Pragmatic managers Who Have Adopted a Less Aggressive Posture and Are Reaching Out to Form New Coalitions." *National Journal* 17 (June 8): 1350–1353.

Statland, B. E. 1992. "Nutrition and Cancer." *Clinical Chemistry* 38 (8B): 1587–1594.

Stern, Paul C., Oran R. Young, and Daniel Druckman, eds. 1992. *Global Environmental Change: Understanding the Human Dimensions.* Washington, D.C.: National Academy Press.

Stevens, Richard G., and Suresh H. Moolgavkar. 1984. "A Cohort Analysis of Lung Cancer and Smoking in British Males." *American Journal of Epidemiology* 119 (4): 624–641.

Stevens, William K. 1990a. "Risk Is Seen in Needless Removal of Asbestos." *New York Times,* January 19, p. A20.

——. 1990b. "Worst Fears about Acid Rain Unrealized." *New York Times,* February 20, pp. C1, C11.

——. 1995a. "Scientists Say Earth's Warming Could Set Off Wide Disruptions." *New York Times*. September 18, pp. A1, A8.

——. 1995b. "In Rain and Temperature Data, New Signs of Global Warming." *New York Times*. September 26, pp. C4.

——. 1996. "At Hot Center of Debate on Global Warming." *New York Times*. August 6, 1996. pp. C1ff.

Stockwell, H. G., A. L. Goldman, G. H. Lyman, C. I. Noss, A. W. Pinkham, E. C. Candelora, and M. R. Brusa. 1992. "Environmental Tobacco Smoke and Lung Cancer Risk in Nonsmoking Women." *Journal of the National Cancer Institute* 84, no. 18 (September 16): 1417–1422.

Stolberg, Sheryl Gay. 1998. "New Cancer Cases Decreasing in U.S. as Deaths Do, Too." *New York Times*, March 13, p. A1.

Stone, Richard. 1995. "Panel Slams EPA's Dioxin Analysis." *Science* 268 (May 26): 1124.

Stretton, Hugh. 1976. *Capitalism, Socialism, and the Environment*. New York: Cambridge University Press.

Strickland, Stephen P. 1972. *Politics, Science, and Dread Disease: A Short History of United States Medical Research Policy*. Cambridge, Mass.: Harvard University Press.

Sugg, Ike C. 1996. "Selling Human Rights Saves Animals." *Wall Street Journal*, July 23, p. A22.

Sugimura, Takashi. 1986. "Studies on Environmental Chemical Carcinogenesis in Japan." *Science* 233, July 18: 312–318.

Sutherland, Fredric P., and Vawter Parker. 1988. "Environmentalists at Law." In Peter Borrelli, ed., *Crossroads: Environmental Priorities for the Future*, 181–190. Washington, D.C.: Island Press.

Tango, T. 1994. "Effect of Air Pollution on Lung Cancer: A Poisson Regression Model Based on Vital Statistics." *Environmental Health Perspectives* 102, supplement 8 (November): 41–45.

Tierney, John. 1996. "Recycling Is Garbage." *New York Times Magazine*, June 30: 24–29ff.

Tobin, Richard J. 1984. "Revising the Clean Air Act: Legislative Failure and Administrative Success." In Norman J. Vig and Michael E. Kraft, eds., *Environmental Policy in the 1980s: Reagan's New Agenda*. Washington, D.C.: Congressional Quarterly Press.

Tolman, Jonathan. 1996. "Real Pests Aren't in the Food." *Wall Street Journal*, September 18, p. A23.

Tredaniel, J., P. Boffetta, R. Saracci, and A. Hirsch. 1994. "Exposure to Environmental Tobacco Smoke and Risk of Lung Cancer: The Epidemiological Evidence." *European Respiratory Journal* 7, no. 10 (October): 1877–1888.

Tucker, William. 1982. *Progress and Privilege: America in the Age of Environmentalism*. Garden City, N.Y.: Anchor Press.

Tufts University Diet and Nutrition Letter. 1996a. "Will Eating Less Fat Lower Breast Cancer Risk After All?" Vol. 14, no. 2 (April): 1–2.

Tufts University Diet and Nutrition Letter, 1996b. "How Much Are Pesticides Hurting Your Health?" Vol. 14, no. 2 (April): 4–6.

Udall, Stewart L. 1988. *The Quiet Crisis and the Next Generation.* Salt Lake City: Peregrine Smith.

United States Air Force. 1995. Draft of report, *Air Force Study and Epidemiologic Investigation of Health Effects in Air Force Personnel Following Exposure to Herbicides.*

Van Houtven, George L., and Maureen L. Cropper. 1994. "When Is a Life Too Costly to Save? The Evidence from Environmental Regulations." *Resources For the Future,* Winter: 6–10.

Van Loon, A. J., J. Burg, R. A. Goldbohm, and P. A. Van den Brandt. 1995. "Differences in Cancer Incidence and Mortality among Socio-Economic Groups." *Scandinavian Journal of Social Medicine* 23 (2): 110–120.

Vig, Norman J., and Michael E. Kraft. 1984. "Environmental Policy from the Seventies to the Eighties." In Norman J. Vig and Michael E. Kraft, eds., *Environmental Policy in the 1980s: Reagan's New Agenda.* Washington, D.C.: Congressional Quarterly Press.

Viscusi, W. Kip. 1992. *Smoking, Making the Risky Decision.* New York: Oxford University Press.

Viscusi, W. Kip, and James T. Hamilton. 1996. "Cleaning up Superfund." *Public Interest* 124 (Summer): 52–60.

Vital STATS: The Numbers Behind the News. 1997. "Is the Sperm Count Down for the Count?" December: 1, 5.

Walljasper, Jay. 1992. "The Anti-Green Backlash." *Utne Reader,* March/April: 158–159.

Wall Street Journal. 1990. "Revenge of the Apples." December 17, p. A8.

Wall Street Journal. 1996. "Nicotine Attack: Cigarette Regulation Is Formally Proposed; Industry Sues to Halt It; Aiming at Use by Children, FDA Says Firms Know the Product Is Addictive; Harsh Government." August 11, p. A1.

Wargo, John. 1996. *Our Children's Toxic Legacy.* New Haven: Yale University Press.

Warren, Karen. 1990. "The Power and Promise of Ecological Feminism." *Environmental Ethics* 12 (Summer): 125–146.

Warren, Karen, and Jim Cheney. 1991. "Ecological Feminism and Ecosystem Ecology." *Hypatia* 6 (1): 179–197.

Washington Post. 1991. "Defamation of a Vegetable." March 22, p. A24.

Washington Post. 1994a. "Study Links Hot Dogs, Cancer: Ingestion by Children Boosts Leukemia Risk, Report Says." June 3, p. A2.

——. 1994b. "Hold the Relish: A Wurst-Case Scenario." June 4, p. G1, G8.

Weiser, Benjamin, 1996. "Big Tobacco vs. Big Media." *Washington Post National Weekly Edition,* January 15–21, 1996, pp 6–9.

Weisskopf, Michael. 1990a. "From Fringe to Political Mainstream." *Washington Post,* April 19, pp. A1, A16.

——. 1990b. "Weary Lawmakers Pass Fiscal Package." *Washington Post,* October 28, p. A1.

Westerstahl, Jorgen, and Folke Johansson. 1991. "Chernobyl and the Nuclear Power Issue in Sweden: Experts, Media, and Public Opinion." *International Journal of Public Opinion Research* 3, no. 2 (Summer): 115–131.

Whelan, Elizabeth. 1978. *Preventing Cancer.* New York: Norton. Rev. ed., 1994. Amherst, New York: Prometheus Books.

——. 1981. "Chemicals and Cancerphobia." *Society* 18, no. 3 (March/April): 5–27.

——. 1992. *Toxic Terror: The Truth behind the Cancer Scare.* Buffalo: Prometheus Books.

——. 1993. "Getting Serious about Non-Risks to Children." *Priorities,* Summer: 5–6.

——. 1994. *Preventing Cancer.* Rev. ed. Amherst, New York: Prometheus Books.

Whitaker, John C. 1976. *Striking a Balance: Environment and Natural Resources Policy in the Nixon-Ford Years.* Washington, D.C.: American Enterprise Institute.

Wildavsky, Aaron. 1979. *Speaking Truth to Power.* Boston: Little, Brown.

——. 1995. *But Is It True?* Cambridge, Mass.: Harvard University Press.

——. 1988. *Searching for Safety.* New Brunswick, N.J.: Transaction Books.

Wildavsky, Aaron, and Karl Dake. 1990. "Theories of Risk Perception: Who Fears What and Why." *Deadalus,* Fall: 41–60.

Williams, James D., and Ronald M. Nowak. 1986. "Vanishing Species in Our Own Backyard: Extinct Fish and Wildlife of the United States and Canada." In Les Kaufman and Kenneth Mallory, eds., *The Last Extinction.* Cambridge, Mass.: MIT Press.

Wilson, Edward O. 1991. "Biodiversity, Prosperity, and Value." In F. Herbert Bormann and Stephen R. Kellert, eds., *Ecology, Economics, and Ethics: The Broken Circle.* New Haven: Yale University Press.

Wilson, James Q. 1980. *The Politics of Regulation.* New York: Basic Books.

——. 1989. *Bureaucracy.* New York: Basic Books.

Wines, Michael. 1983. "Scandals at EPA May Have Done In Reagan's Move to Ease Cancer Controls." *National Journal,* June 18: 1267–1269.

Wise, Jacqui. 1998. "Turnaround in Cancer Trends Seen in the U.S." *British Medical Journal,* vol. 316, no. 7135 (March 21): 886.

Wolpert, Lewis. 1993. *The Unnatural Nature of Science.* Cambridge, Mass.: Harvard University Press.

World Commission on Environment and Development. 1987. *Our Common Future*. Oxford: Oxford University Press.

Wynder, Ernst L., and Gio B. Gori. 1977. "Contribution of the Environment to Cancer Incidence: An Epidemiologic Exercise." *Journal of the National Cancer Institute* 58 (4): 825–832.

Yablokov, Alexey V. 1993. "Russia's Environmental Legacy." *Environmental Science and Technology* 27 (April): 579.

Yamasaki, H., A. Loktionov, and L. Tomatis. 1992. "Perinatal and Multigenerational Effect of Carcinogens: Possible Contribution to Determination of Cancer Susceptibility." *Environmental Health Perspectives* 98 (November): 39–43.

Zeckhauser, Richard, and W. Kip Viscusi. 1990. "Risk within Reason." *Science* 248 (May 4): 559–564.

Ziegler, Charles E. 1987. *Environmental Policy in the USSR*. [Revised, ed. 1990] Amherst: University of Massachusetts Press.

Zimmerman, Michael E., 1994. *Contesting Earth's Future: Radical Ecology and Post Modernity*. Berkeley and Los Angeles: University of California Press.

Index

Environmental Cancer—A Political Disease?

S. ROBERT LICHTER AND STANLEY ROTHMAN

Media reports on environmental cancer are frequent and frightening. Public policy—and public spending—reflect widespread concern over the presence of carcinogens in our air and water and food. Yet how reliable is mass media information about environmental cancer? How accurate are the risk assessments that underlie our public policy decisions?

In this provocative book, S. Robert Lichter and Stanley Rothman examine the controversies surrounding environmental cancer and place them in historical perspective. Then, drawing on surveys of cancer researchers and environmental activists, they reveal that there are sharp differences between the two groups' viewpoints on environmental cancer. Despite these differences, a further comparison—between the views of the two groups and the content of television and newspaper accounts over a two-decade period—shows that press reports most frequently cite the views of environmental activists as if they were the views of the scientific community. These findings cast doubt on the objectivity of the news media and environmental activists. And, the authors conclude, misplaced fears about the risks of environmental cancer have seriously distorted public policy and priorities.

"A fascinating and provocative book that is bound to generate interest, controversy, and debate."—John D. Graham, director, Harvard Center for Risk Analysis

"A fascinating overview of how the U.S. came to transfer enormous resources to hypothetical health risks based on assumptions that have proven to be wrong. The expense is still mounting."
—Bruce N. Ames, University of California

S. Robert Lichter president of the Center for Media and Public Affairs in Washington, D.C., and adjunct professor of political science at Georgetown University. He is the author of many books and articles on the role of news and entertainment media in American society. Stanley Rothman is Mary Huggins Gamble Professor of Government Emeritus and director of the Center for the Study of Social and Political Change at Smith College. His numerous books include *American Elites*, published by Yale University Press.